D0564905

Lifelines

LIFE BEYOND
THE GENE

STEVEN ROSE

OXFORD
UNIVERSITY PRESS

Oxford New York
Auckland Bangkok Buenos Aires Cape Town Chennai
Dar es Salaam Delhi Hong Kong Istanbul Karachi Kolkata
Kuala Lumpur Madrid Melbourne Mexico City Mumbai Nairobi
São Paulo Shanghai Taipei Tokyo Toronto

Published by Oxford University Press, Inc.
198 Madison Avenue, New York, New York 10016

www.oup.com

First published as a paperback by Penguin Press in 1998

First published as an Oxford University Press paperback in 2003

Library of Congress Cataloging-in-Publication Data
Rose, Steven P. R. (Steven Peter Russell), 1938–
Lifelines : life beyond the gene / Steven Rose.
p. cm.
Previously published as: Lifelines : biology beyond determinism.
Includes bibliographical references and index.
ISBN 0-19-515039-2 (pbk.)
1. Biology—Philosophy. I. Title.
QH331 .R645 2001
570'.1—dc21 2001036974

1 3 5 7 9 8 6 4 2

Printed in the United States of America
on acid-free paper

Contents

Preface to the Penguin edition

I have taken the opportunity offered by the paperback edition of the book to correct some stylistic infelicities and clarify some excessively obscure sentences and ambiguities, as well as to update some of the footnotes. I would also like to acknowledge the helpful points made particularly by two reviewers of the hardback. Jon Turney detects in my writing a dislike of metaphor, and it is true that I am very critical of how metaphors may mislead if they are confused with analogy or homology. However, it is also the case that some of the most powerful theoretical advances in science have taken off from inspired metaphors. Like alcohol, they can be joyously intoxicating, but in excess you risk a bad hangover. Tim Ingold points out that I toss the term history around over-casually, and run the danger of conflating evolutionary, developmental and social historical processes. Of course these are very different; even their time axes involve different scales, and temporal processes in general get shorter shrift in my chapters than I had originally intended – a theme I might return to in some later book.

Steven Rose, February 1998

Preface

The rise of the present enthusiasms for biologically determinist accounts of the human condition date to the late 1960s. They were not initiated by any specific advance in biological science, or powerful new theory, but harked back instead to an earlier tradition of eugenic thinking which, still strong especially in the USA during the 1930s, had been eclipsed and driven into intellectual and political disrepute in the aftermath of the war against Nazi Germany and its racially inspired Holocaust. A series of UNESCO-sponsored statements by geneticists, anthropologists and social scientists, which followed the end of that war, spelled out what became the consensus view for the next quarter-century, that the roots of human inequality lay not so much in the uniqueness of our genes as in the unequal distribution of wealth and power between nations, races and classes (the question of gender inequality was never raised by these consensual groups).

The 1960s, that decade of hope for humanity, saw struggles for social justice across the globe; the rise of great movements of national, black and then women's liberation, catalysed, especially in the industrialized countries, by students. In response, as it were, to these movements came the reassertion of old but hitherto submerged claims: that on average black and working-class intelligence was genetically inferior to that of whites and the middle class, and that patriarchal domination was an inevitable consequence of genetic and hormonal differences between men and women. Initially such claims drew on no new research, but instead warmed over older traditions in biological and psychological thinking. It was not until the mid-1970s, with the emergence of a new and more grandiose set of theories, described as sociobiology, that the biologically determinist viewpoint became more theoretically coherent. Its position could be encapsulated in the sound-bite phrase 'the selfish gene', a view I characterize in this book as 'ultra-Darwinism'.

Such assertions were energetically contested by many biologists and social scientists, especially those of us who had aligned ourselves with what in those more optimistic days was described as the radical science movement. The grounds for our opposition were both scientific and political. Ultra-Darwinism and sociobiological theorizing, especially as applied to human societies, rested on shaky empirical evidence, flawed premises and unexamined ideological presuppositions concerning so-called universal aspects of 'human nature'. Furthermore, such determinist claims were immediately marshalled in support of neo-Fascist and New Right political movements across the USA, Britain and continental Europe. It was in this context that the sociologist Hilary Rose and I edited a series of books (*The Political Economy of Science* and *The Radicalisation of Science* in the mid-1970s, and *Against Biological Determinism* and *Towards a Liberatory Biology* in the early 1980s), and in the mid-1980s the geneticist Dick Lewontin, psychologist Leo Kamin and I wrote *Not in Our Genes*, intended as a comprehensive attempt to analyse and contest both the ideology and scientific claims of biological determinism.

These of course were far from the only rebuttals in what became something of a battle of the books. But in the last decade, especially in the context of dramatic advances in the sciences of both genes and brains, the stream of ultra-Darwinist and biologically determinist claims has become a torrent. First the Human Genome Project, the major international effort to map and then sequence all human genes, and then the Decade of the Brain (more than halfway through in the USA as I write, barely started in Europe) have not merely offered the possibility of vastly increasing our knowledge of aspects of human biology, but have also held out the promise of further and further technological power to manipulate both genes and minds in the interests both of individual health and of greater social tranquillity. Techniques of intervention barely imaginable a decade ago, or at best the stuff of science fiction, now rate stock market quotations and turn academic researchers into entrepreneurial millionaires.

To judge from headlines in daily newspapers, or the titles of academic papers in major scientific journals, the issues of a decade ago have been settled. Vulgar sociobiology may be out, but what I have called neurogenetic determinism is strongly entrenched. There are genes available to account for every aspect of our lives, from personal success to existential despair: genes for health and illness, genes for criminality, violence and 'abnormal' sexual orientation – even for 'compulsive shopping'. And genes too to explain, as ever, the social inequalities that divide our lives along lines of class, gender, race, ethnicity . . . And where there are genes, genetic and pharmacological engineering hold out hopes for salvation that social engineering and politics have abandoned.

The challenge to the opponents of biological determinism is that, while we may have been effective in our critique of its reductionist claims, we have failed to offer a coherent alternative framework within which to interpret living processes. We may reply, with some justification, that we have been too busy attempting to rebut the determinists, but sooner or later it becomes necessary for us to fight fire with fire, to try to spell out more coherently our contrasting biological case. *Lifelines* originated as an attempt to meet that challenge. Shortly after my previous book, *The Making of Memory*, had appeared, my then editor at Penguin, Ravi Mirchandani, suggested that the time was ripe for a book on the philosophy of biology, not from the perspective of a professional philosopher but from that of someone who, like myself, is both an experimental biologist with an ongoing laboratory commitment, and someone concerned with both the theory and social framing of my science. John Brockman, my agent, as indeed he is of several of those whose positions I strongly criticize in this book (but then John enjoys acting as impresario to scientific debates), helped shape my early structural ideas for the book.

I have tried to achieve a number of goals: first, to convey what it means to 'think like a biologist' about the nature of living processes; second, to analyse both the strengths and the limitations of the reductionist tradition which dominates much of biology; and third, to offer a perspective on biology which transcends genetic reductionism, by placing the organism, rather than the gene, at the centre of life – this is the perspective that I call *homeodynamic*. To arrive at these goals, I have had to try to understand the historical roots of current biological thought, and draw upon those powerful alternative traditions in biology which have refused to be swept along by the ultra-Darwinist tide into accepting that living processes can be reduced to mere assemblages of molecules driven by the selfish urges of the genes to make copies of themselves. These traditions argue instead the need for a more holistic, integrative biology, one which understands and enjoys complexity and recognizes the need for epistemological diversity in our explorations of the nature and meaning of life. Their voices can still be heard above the ultra-Darwinist din.

Furthermore, in order to stress the positive case that I have wished to make, in places I have had to set it against the opposite view as presented in its rhetorically strongest form. To do so, I have had to choose appropriate foils. The two authors who have served me best in this respect are the sociobiologist Richard Dawkins, whose several books speak with a single ultra-Darwinist voice, and the philosopher Daniel Dennett, whose *Darwin's Dangerous Idea* carries ultra-Darwinism to the furthest reaches. Among practising biologists – those who spend a significant part of every working day thinking about and designing experiments, persuading some research body to fund them and then actually carrying them out in the laboratory – there is an audible grumbling about why 'we' should give the claims of either Dawkins or Dennett serious consideration. These are, after all, people who either no longer do science or never did it; they are not part of 'our' discourse of careful experimentation and allied theoretical claim. Yet this professional complaint, often made by colleagues I deeply respect, misses the point. Dawkins, Dennett and their camp-followers, as best-selling authors in the public understanding of science lists, frame the public debate. We can see their influence on the writers and readers of Sunday newspapers, and on politicians and novelists alike. Culturally, they are too important for practising biologists to ignore them. Here I criticize many of their arguments robustly; but it is the arguments, together with the metaphysical assumptions behind the arguments and their implications for both biology and culture, that concern me, not the individuals who put them forward. The stakes are high: how do we, not just as biologists but as denizens of the late twentieth century, culturally understand nature?

One further point of clarification is necessary. In attacking ultra-Darwinism in this way I want to make it absolutely clear that I have no intention of departing from a materialist view of life, nor of giving any ground at all to anti-Darwinian fundamentalists, creationists or New Age mystics of any shape or hue. I view the world from a strongly materialist perspective – one, however, which stresses both ontological unity and epistemological diversity – a position I also tried to spell out in *The Making of Memory*. So far as possible, *Lifelines*, like the memory book but unlike *Not in Our Genes*, is a within-biology discussion. That is, I largely refrain from discussing the ideology, social origins or social consequences of ultra-Darwinism and reductionism. However, it would not have been either possible or proper to leave these issues entirely unaddressed, and I have tried to summarize them in the penultimate chapter, 'The Poverty of Reductionism'. This itself is built around an analysis I first published as a 'Commentary' article in *Nature* in 1995 as 'The rise of neurogenetic determinism', an extended version of which appeared later that year in the second issue of the new journal *Soundings*.

In writing this book, I have incurred a large number of intellectual debts. Dick Levins and Dick Lewontin, in their book of essays *The Dialectical Biologist*, and more recently Levins and Yrjo Haila, in *Nature and Humanity*, helped provide the theoretical framework which informs my text. So too from rather different perspectives have Brian Goodwin (*How the Leopard Changed Its Spots*) and Mae-Wan Ho (*The Worm and the Rainbow*). I have learned from all these books and their authors, and also from Stuart Kauffman's chaos-theory approach to biology, *At Home in the Universe*, and Hilary Rose's *Love, Power and Knowledge*. Brian Goodwin also kindly made available to me the pre-publication manuscript of his and Gerry Webster's *Form and Transformation*, though he will not, I know, be happy with my dismissal in Chapter 2 of natural kinds in biology.

Apart from my *Nature* article, some of the ideas and themes of the book have been tested at seminars and discussion groups over the period of writing, notably at a Nobel Forum symposium at the Karolinska Institute in Stockholm in January 1996, the 1996 Spoleto-Scienza and Edinburgh International Science Festivals, and Open University summer schools. Chairing the Open University course 'Living Processes' during the gestation of the book between 1993 and 1995 also helped sharpen some of my thoughts and arguments. I am grateful for the hospitality of Aant Elzinger's Department of Science Theory, at the University of Goteborg, during October and November 1995 while drafting several of the chapters. I must also thank colleagues, visitors and students in the Brain and Behaviour Research Group and Biology Department at the Open University for their indulgence at times when, over the past couple of years, my thoughts have strayed from the immediate experimental tasks in hand to the more general issues covered here.

Discussions spread over two continents and many years with Enrico Alleva, Kostya Anokhin, Giorgio Bignami, Ruth Hubbard, Dick Levins, Dick Lewontin, Radmila (Buca) Mileusnic, Luciano Terrenato and Ethel Tobach are also reflected in many of the arguments that follow. Several people have read and commented on earlier drafts of the whole book or of individual chapters, and I am particularly grateful to Rusiko Burchuladze, Brian Goodwin, Ruth Hubbard, Charles Jencks, Hilary Rose, Jonathan Silvertown, Miroslav Simic, Lars Terenius and Pat Wall, and to several anonymous reviewers, for correcting errors, helping strengthen arguments and putting me right in places where I had gone off the rails. That great biochemist and scholar N. W. (Bill) Pirie read the manuscript and made detailed cover-to-cover comments – perhaps his last intellectual interventions before he died in March 1997, aged 89 and still working in his lab virtually till the day of his death. I shall greatly miss his crusty wisdom and advice. John Woodruff, dedicated subeditor, went beyond the call of duty in clarifying obscurities in my prose – and hence in my thinking. A special thanks, too, to Renate Prince for providing architectural and historical advice and source material which enabled me to deal with Dennett's arguments about spandrels and adaptationism in Chapter 8. As throughout the past thirty-five years, my huge debt to the continued dialectic of discussion (to say nothing of love) with Hilary Rose remains irreducible to figures or even words.

None of the above-mentioned people should necessarily be assumed to agree with every argument in the book – and, of course, I am solely responsible for such errors as remain.

London, February 1997

Credits

Figure 2.1(a) from a photograph by R. C. James.
Figure 2.1(b) from *The Brain*, Christine Temple, Penguin, 1993.
Figure 2.2(b) from *The Chemistry of Life*, Steven Rose, Penguin, 1966.
Figure 3.1 from *Anatomia et Contemplatio*, A. de Leewenhoek, 1685.
Figure 3.2 from *Microscopical Researches into the Accordance in the Structure and Growth of Animals and Plants*, 1847.
Figure 3.3 courtesy of Heather Davies, Open University.
Figure 3.4 from Rosalind Franklin and Ray Gosling, *Nature*, 171, 740, 1953. Reprinted with permission from Macmillan Magazines Ltd.
Figure 3.5 courtesy of Dr Radmila Mileusnic, Open University.
Figure 4.1 redrawn by Nigel Andrews from *The Chemistry of Life*, Steven Rose.
Figure 4.2(a) courtesy of Dr Michael Stewart, Open University.
Figure 4.4 redrawn by Nigel Andrews after Arthur Koestler in *Beyond Reductionism*, ed. A. Koestler and J. R. Smythies, Hutchinson, 1969.
Figure 5.2 drawn by Nigel Andrews after Lewis Wolpert, *The Triumph of the Embryo*, Oxford University Press, 1991.
Figure 5.6 from *The Chemistry of Life*, Steven Rose.
Figure 6.3 modified and redrawn after Irwin B. Levitan and Leonard K. Kaczmarek, *The Neuron: Cell and Molecular Biology*, Oxford University Press, 1991.
Figure 6.4 courtesy of Dr Luigi Aloe, Institute of Neurobiology, CNR, Rome.
Figure 6.8 from *At Home in the Universe*, Stuart Kauffman, Viking, 1995.

Figure 6.9 reprinted with permission from James Lechleiter, Steven Girard, Ernest Peralta and David Clapham, *Science*, **252**, 124, 1991, copyright American Association for the Advancement of Science.

Figure 6.10(a) reprinted with permission from *The Molecular Biology of the Cell*, Albert et al., copyright Garland Publishing Inc.

Figure 7.1(a) from *The History of Creation*, Ernst Haeckel, New York, 1879.

Figure 7.1(b) from *Introduction to the Study of Man*, J. Z. Young, 1971 by permission of Oxford University Press.

Figure 7.2 from *Journal of Researches into the Geology and Natural History of the Countries Visited during the Voyage of HMS Beagle*, Charles Darwin, reprinted London, 1891 (reproduced by kind permission of the Mary Evans Picture Library).

Figure 8.2(a) and 8.2(b) from *Wonderful Life*, Stephen Jay Gould, Viking, 1989.

Figure 8.3 from *Kunstformen der Natur*, Ernst Haeckel, Leipzig, 1904.

Figure 8.5 from *On Growth and Form*, D'Arcy Thompson, abridged edition, 1961, by permission of Cambridge University Press.

Figure 8.6(a) courtesy of Mike Levers, Open University.

Figure 8.6(b) reproduced by kind permission of the Science Photo Library.

Figure 9.1 from *At Home in the Universe*, Stuart Kauffman, Viking, 1995.

Figure 9.2 courtesy of Dr David S. McKay, NASA/JSC, Houston, Texas.

Figures 9.3(a), 9.3(b) and 9.3(c) reprinted with permission from Stephen Mann and Geoffrey A. Ozin, *Nature*, **382**, pp 313–317, 1996, copyright Macmillan Magazines Ltd.

Figure 9.5 from *At Home in the Universe*, Stuart Kauffman, Viking, 1995.

Figures, 1.1, 1.2, 1.3, 2.2(a), 2.2(c), 4.2(b), 4.3, 5.1, 5.3, 5.4, 5.5, 6.1(b), 6.2, 6.3, 6.5, 6.6, 6.7, 8.1, 9.4, 10.1 and 10.2 all drawn by Nigel Andrews.

All other figures supplied by the author.

I

Biology, Freedom, Determinism

Man first of all exists, encounters himself, surges up in the world – and defines himself afterwards . . . he will be what he makes of himself. Thus there is no human nature . . . Man simply is. He is what he wills . . . One will never be able to explain one's action by reference to a given and specific human nature – in other words there is no determinism: man is free, man is freedom.

Jean-Paul Sartre, *Existentialism and Humanism*[1]

We are survival machines – robot vehicles blindly programmed to preserve the selfish molecules known as genes.

Richard Dawkins, *The Selfish Gene*

LIFE ITSELF

A new baby stares gravely up at her mother and her entire face curls into an unmistakable smile.

Spring, and the sticky yellow and green horse-chestnut buds slowly unfurl. Courting birds flit between the trees.

Summer, and clouds of small black midges surround us as we walk the moors.

Autumn, and amid the fallen leaves of the beech wood a miniature forest of mushrooms sprouts.

An African plain: termite mounds rise skywards, inhabited by hundreds of scurrying thousands.

A coral reef: myriads of brightly striped and patterned fish dart in and out of crevices; shoals weave and turn, each individual effortlessly part of the choreographed unity of the greater whole.

A drop of pondwater: single-celled, almost transparent creatures ooze; occasionally one meets and engulfs another.

All alive. All making their individual and collective ways in the world, cooperating, competing, avoiding, living with, living off, interdependent. All the present-day products of some four billion years of evolution, of the continued working-out of the great natural experiments that the physical and chemical conditions of planet Earth have made possible, perhaps inevitable. For every organism, a lifeline – its own unique trajectory in time and space, from birth to death.

The sheer scale, diversity and volume of life on Earth surpasses the imagination. Take a square metre of European or North American forest and slice off the top 15 centimetres of soil, and you will find, among numerous other life forms, as many as 6 million tiny worms – nematodes – perhaps 200 different species. It is possible that there are as many as 10,000 species of bacterium in a single gram of soil, yet only 3,000 have so far been identified and named by microbiologists. Conservative estimates put the number of different species on Earth at 14 million; no one knows for sure and some have claimed that there are at least 30 million. Of these, only a few per cent – 2 million at most – have been studied, identified, named. Indeed, almost all biological research has been based on a few hundred different life forms at most. The smallest independently living organisms are no more than 0.2 micrometres – one-fifth of a millionth of a metre – in diameter; the largest living animal, the blue whale, can grow to more than 30 metres and may weigh 200 tonnes – heavier than any known extinct dinosaur. Bacteria live for 20 minutes or so before dividing into two; near where I live in Yorkshire is an elderly oak tree which was noted in William the Conqueror's Domesday Book nearly a thousand years ago. And some Californian redwoods far outstrip whales and oak trees, reaching nearly 100 metres in height and at least 2,500 years of age.

What a world to be living in, to marvel at, to enjoy in all its multifarious variety. 'O brave new world, that has such creatures in't,' to paraphrase the old wizard Prospero's daughter Miranda in

The Tempest. And her voice echoes the feelings of poets, painters and writers throughout recorded history.

But to study, to interpret, to understand, to explain and to predict? These are the tasks of myth-makers, magicians and, above all today, of scientists, of biologists. I am of this last category. We seek not to lose the visions provided by writers and artists, but to add to them new visions which come from the ways of knowing that biology, the science of life, opens up. These ways can show beauty also below the surface of things: in the scanning electron microscope's view of the eye of a bluebottle as much as in the flowering of a camellia; in the biochemical mechanisms that generate usable energy in the minuscule sausage-shaped mitochondria that inhabit each of our body cells, as much as in the flowing muscles of the athlete who exploits these mechanisms.

How are we to understand these multitudes of organisms, these orders-of-magnitude differences in space and time encompassed by the common definition of living forms? Humans are like, yet unlike, any other species on Earth. We have had to learn to adapt to, domesticate, subordinate, protect ourselves from or exist harmoniously with a goodly proportion of the other creatures with which we share our planet. And in doing so, to make theories about them. Every society that anthropologists have studied has developed its own theories and legends to account for life and our place within it, to interpret the great transitions that characterize our existence; the creation of new life at birth and its termination at death. In most societies' creation myths, a deity imposes order upon the confused mass of struggling life. Although our own society is no exception, we now phrase things differently, claiming to have transcended myth and replaced it with secure knowledge. For the last three hundred years, Western societies have built on and transcended their own creation myths by means of *scientia*, the organized investigation of the universe, made possible within the rules and by the experimental methods of natural science, and with the aid of powerful instruments designed to extend the human senses of touch, smell, taste, sight and sound.

THE POWER OF BIOLOGY

The power of Western science as it has developed over the past three centuries derived in the first instance from its capacity to explain, and later from its power to control, aspects of the non-living world in the province of physics and chemistry. Only subsequently were the methods and theories that had been shaped by the success of these older sciences turned towards the study of living processes. The several sciences that today comprise biology have been barely six generations in the making, and have been transformed beyond measure even within my own lifetime. Despite our ignorance of the overwhelming majority of life forms which exist on Earth today (indeed, most biochemical and genetic generalizations are still derived from just three organisms: the rat, the fruit fly and the common gut bug *Escherichia coli*), and our inability to do more than offer informed speculations about the processes that have given rise to them over the past 4 billion years, we biologists are beginning to lay claims to universal knowledge, of what life is, how it emerged and how it works. In all life forms, in all living processes, we argue, certain general principles hold; certain mechanisms, certain forms of chemistry, exist in common. Some have even gone further, arguing that what they deduce to be true of life on Earth is but a special case of a phenomenon so universal that its rules must apply to all living forms anywhere in the universe.

The successes of science have been based not so much on observation and contemplation, but on active intervention in the phenomena for which explanations were being sought. When addressed to purely chemical and physical processes, such interventions seldom present significant problems of ethics, of challenges to the very right of the researcher to intervene. But there is no doubt that intervention in living processes confronts us all – not just researchers, but also the society which has come to depend upon the results of their research – with moral dilemmas. We cannot escape the fact that interventionist biology, and above all physiology, is a science built on violence, on 'murdering to dissect', and that hitherto there has been no alternative means of discovering the intimate molecular and cellular events that, at least on one level of description, constitute life itself. The reductive philosophy that has proved so seductive to biologists yet so hazardous in its consequences seems an almost inevitable product of this interventionist and necessarily violent methodology.

More than most sciences today, biology impinges directly on how we live. Like chemistry and physics, its technologies transform our personal, social and natural environments via pharmacology, genetic engineering and agribusiness. Biology also makes claims as to who we are, about the forces that shape the deepest aspects of our personalities, and even about our purposes here on Earth. The claims of the science have become so strong as to seem no longer a matter for debate: they are now the natural way to view the living world. Indeed, today we even use the name given to the science, *biology*, to replace its field of study – life itself and the processes which sustain it; the science has usurped its subject. So 'biological' becomes the antonym not for 'sociological' but for 'social'.[2]

FREEDOM AND DETERMINISM

Hence the epigraphs to this chapter. These two diametrically opposed views on the nature of human nature, of the relationship between our thoughts and actions on the one hand and our chemical constitution – DNA's way of making more DNA – on the other, represent the extremes between which I have tried to steer this book. The first, a windily rhetorical paean to the dignity of universalistic man (I suspect the gendering is not irrelevant) written just after the liberation of France from Nazi occupation, is from the existentialist philosopher Jean-Paul Sartre. The second, with all the brash style of a cheeky adolescent cocking a snook at everything his elders hold dear, is from Richard Dawkins, the St John the Baptist of sociobiology, and was drafted in the comfort of an Oxford college in the mid-1970s. Each has been fashionable in its time, but there is no doubt which better reflects the spirit of the past two decades.

Each of course is more an exercise in political sloganeering than a sustainable philosophical position. How does Sartre's freedom deal with the inexorability of human decline, the ravages of cancer, the destructive onset of Alzheimer's disease? And how does Dawkins' gene's-eye view of the world account for the horrors of the Nazi concentration camps or the heroes and heroines of the French resistance? Of course, neither viewpoint sprang fully formed from its author's pen; each had a long lineage in religious, philosophical and scientific debate. And I am not so naïve as to assume that my argument with regard to both positions will be the last word on the subject. However, it is worth stating my thesis right from the beginning. Humans are not empty organisms, free spirits constrained only by the limits of our imaginations or, more prosaically, by the social and economic determinants within which we live, think and act. Nor are we reducible to 'nothing but' machines for the replication of our DNA. We are, rather, the products of the constant dialectic between 'the biological' and 'the social' through which humans have evolved, history has been made and we as individuals have developed (and note already in this sentence my elision of the science of biology with the subject of its study, human life).

To argue otherwise is fundamentally to misunderstand the nature of living processes which it is the purpose of biological science to identify and interpret. Furthermore, our difficulty in thinking our way beyond such antitheses, often expressed as a false dichotomy between nature and nurture, itself derives from the social, philosophical and religious framework within which modern science has developed since its origins, contemporaneous with the birth of capitalism, in seventeenth-century north-western Europe. But it is as a biologist by training and trade, rather than as a philosopher or historian of science, that I shall argue that the naïve – even vulgar – reductionism and determinism which often masquerade as representative of how biology perceives the world is mistaken. It is not that we are the isolated, autonomous units of Sartre's imagination; rather, our freedom is inherent in the living processes that constitute us.

The science we do, the theories we prefer, and the technologies we use and create as part of that science can never be divorced from the social context in which they are created, the purposes of those who fund the science, and the world-views within which we seek and find appropriate answers to the great *what*, *why* and *how* questions that frame our understanding of life's purposes. So, certainly, with modern biology, whose multifarious answers to these questions are imbued with social and political significance. The prevailing fashion for giving genetic explanations to account for many if not all aspects of the human social condition – from the social inequalities of race, gender and class to individual propensities such as sexual orientation, use of drugs or alcohol, or the failures of the homeless or psychologically distressed to survive effectively in modern society – is the ideology of *biological determinism*, typified by the extrapolations of evolutionary theory that comprise much of what has become known as *sociobiology*.[3] (This is the assemblage of theories and assertions about humans and society which claims that it is evolutionary theory rather than sociology, economics or psychology that can best explain how and why we live as we do.) It is not possible to write a book such as this without referring to these claims and their politics, and I shall certainly question their legitimacy. But this is not my main task. It is rather to offer an alternative vision of living systems, a vision which recognizes the power and role of genes without subscribing to genetic determinism, and which recaptures an understanding of living organisms and their trajectories through time and space as lying at the centre of biology. It is these trajectories that I call *lifelines*. Far from being determined, or needing to invoke some non-material concept of free will to help us escape the determinist trap, it is in the nature of living systems to be radically indeterminate, to continually construct their – our – own futures, albeit in circumstances not of our own choosing.

THE COMPLEXITY OF BIOLOGY

Science is assumed to be about both explaining and predicting. There is commonly supposed to be a hierarchy of the sciences, from physics through chemistry, biology and the human sciences. In this scheme physics is seen as the most fundamental of the sciences. There are several reasons for this. Partly, physics is believed to deal with the most general principles by which nature is organized. It both provides explanations of natural phenomena and predicts outcomes, from the falling of an apple to an eclipse of the Moon. Furthermore, the 'laws' of physics apply to biology, but if there are 'laws' of biology they do not apply to non-living systems. Physics is thus a 'hard' science, whose principles can be expressed mathematically, and so it is supposed to be the model to which all other sciences should aspire. By contrast, the social and human sciences are seen as the 'softest' because they are the least capable of precise mathematical expression, and because they do not neatly fit the definition of what 'science' is about set out in the first sentence of this paragraph. Indeed, it can be argued that the 'predictive' tag is put there precisely to privilege simple sciences like physics and chemistry, which were the first of the modern sciences to develop, against those, like the social sciences and many areas of biology, which (as will become clear in what follows) are multiply determined, and do not even set out to predict (Figure 1.1).

For many, scientists and lay public alike, the hierarchical convention none the less seems obvious, natural. Early in the twentieth century there was a determined effort by physicists and philosophers to insist on a unity of the sciences in which, in due course, physics would triumph. Orthodox philosophy is still mainly a philosophy of physics premised on the reductionist view that the task of science is ultimately to collapse biology into chemistry and chemistry into physics, deriving a limited number of universal laws which will explain the entire universe. The physicist Steven Weinberg has argued this reductionist case with elegance and passion in his book *Dreams of a Final Theory*.[4] He takes care to point out that many biologists will not concede such reductionism, recounting his own disagreements with the evolutionary biologist Ernst Mayr.[5] But Weinberg's view remains popular. 'There is only one science, physics: everything else is social work' as molecular biologist James Watson has put it with characteristic robustness.[6] And many biologists, whose own experimental programmes should perhaps help them know better, accede willingly.[7]

Yet there is nothing inevitable about such a hierarchical view. It is a historically determined convention which reflects the particular traditions of the ways in which Western science has developed from its origins in the seventeenth century. Physics deals with relatively simple, reproducible phenomena which can be measured with exquis-

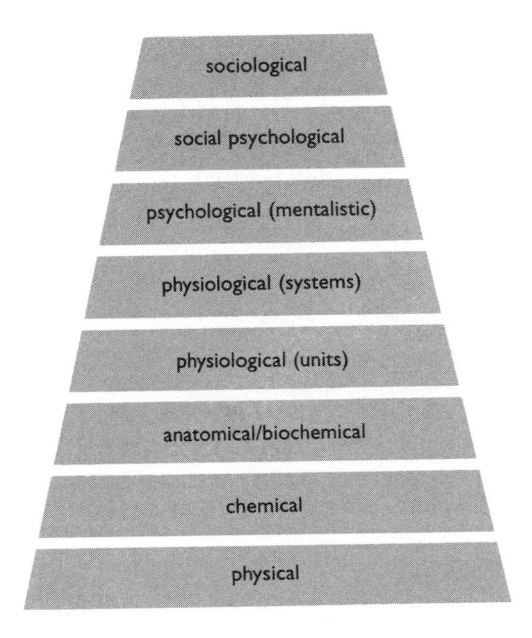

Figure 1.1 The traditional hierarchy of the sciences.

ite precision, and finds it hard to deal with complexity. Biologists' questions about the world are not easily answerable in the reduced, mathematicizing language of physics, and they are said to suffer from a sense of inferiority, of 'physics envy' (which may perhaps be why these days many molecular biologists try to behave as if they *are* physicists!). But we should not be afraid to cut ourselves loose from the reductionist claims that there is only one epistemology, one way to study and understand the world, one science, whose name is physics. Not everything is capable of being captured in a mathematical formula. Some properties of living systems are not quantifiable, and attempts to put numbers on them produce only mystification (as, for instance, with attempts to score intelligence or aggression, or calculate how many bits of information – memories – the brain can store). Biology needs to be able to declare its independence from spurious attempts to mathematicize it. To see why, here's a fable:

FIVE WAYS OF LOOKING AT FROGS

Once upon a time, five biologists were having a picnic by a pool, when they noticed a frog, sitting on the edge, suddenly jump into the water (Figure 1.2). A discussion began: why did the frog jump?

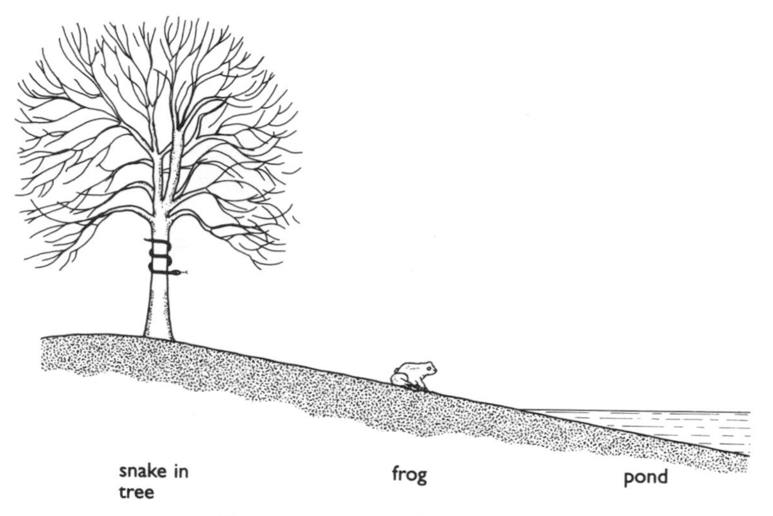

Figure 1.2 Frog, snake and pond.

Says the first biologist, a physiologist, 'It's really quite straightforward. The frog jumps *because* the muscles in its legs contract; in turn these contract because of impulses in the motor nerves arriving at the muscles from the frog's brain; these impulses originate in the brain because previous impulses, arriving at the brain from the frog's retina, have signalled the presence of a predatory snake.'

This is a simple 'within-level' causal chain: *first* the retinal image of the snake; *then* the signals to the brain; *then* the impulses down the nerves from the brain; *then* the muscle contraction – one event following the other, all in a few thousandths of a second (Figure 1.3). Working out the details of such causal sequences is the task of physiology.

'But this is a very limited explanation,' says the second, who is an

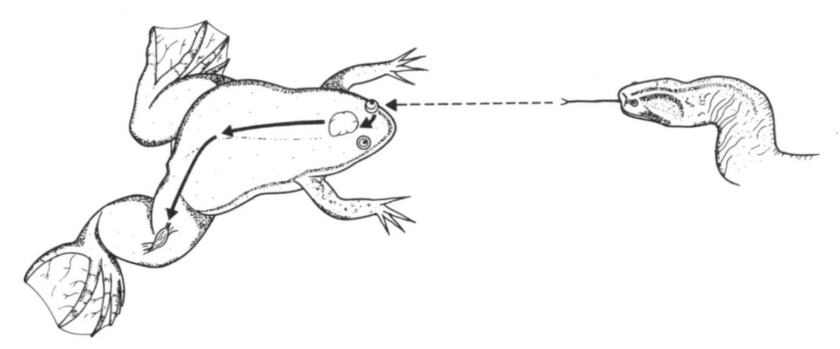

Figure 1.3 What causes the frog to jump?

ethologist, and studies animal behaviour. 'The physiologist has missed the point, and has told us *how* the frog jumped but not *why* it jumped. The reason *why* is *because* it sees the snake and *in order to* avoid it. The contraction of the frog's muscles is but one aspect of a complex process, and must be understood in terms of the goals of that process – in this case, to escape being eaten. The ultimate goal of avoiding the snake is essential to understanding the action.'

Such goal-directed explanations, which are known as *teleonomic*, have given more trouble to philosophers than almost anything else in biology; they are sometimes regarded as bad form, yet they make more everyday sense than most other explanations.[8] They insist that an organism, a piece of behaviour or of physiology, can be understood only within an environmental context which includes both its physical surroundings and other living, socially interacting neighbours. (Indeed, when the organism is a member of that very peculiar species, *Homo sapiens*, then further complexities, those of personal and collective history, come strongly into play.) This type of explanation is a 'top-down' one (it is sometimes called a *holistic* explanation, a dangerously ambiguous word, which I shall avoid). But notice that, unlike the physiologist's explanation, it is not causal in the sense of describing a temporal chain of events in which first one thing, the nerve firing, and then another, the muscle contraction, happen one after the other in time. The jump inevitably precedes achieving the goal towards which it is directed. Thus when animal behaviourists – ethologists –

talk of causes, they do so quite differently from physiologists.

'Neither the physiologist's nor the ethologist's explanations are adequate,' says the third biologist, who studies development. 'For the developmentalist, the only reason that the frog can jump at all is *because* during its development, from single fertilized egg through tadpole to mature animal, its nerves, brain and muscles have become "wired up" in such a way that such sequences of activity are inevitable – or at least, the most probable given any set of starting conditions.'

The process of wiring is an aspect of *ontogeny*, the development of the organism from conception to adulthood, and is addressed by genetics and developmental biology. Unlike the first two explanations, the ontogenetic approach introduces a historical element into the account: the individual history of the frog becomes the key to understanding its present behaviour. Ontogeny is often seen as a dialogue – even a dichotomy – between nature (genetics) and nurture (environment). There have even been attempts to mathematicize this split, and to ask how much is contributed by genes and how much by environment. As will become clear in later chapters, this is a spurious dichotomy and I shall endeavour to transcend it.

'None of these three explanations is very satisfactory,' counters the fourth biologist, an evolutionist. 'The frog jumps *because* during its evolutionary history it was adaptive for its ancestors to do so at the sight of a snake; those ancestors that failed so to do were eaten, and hence their progeny failed to be selected.'

This type of explanation presents problems of defining just what is meant by terms like 'adaptive' and 'selected', problems which have been raised most sharply in the polemical debate over sociobiology, and which I shall examine rather critically in later chapters. One might contrast the developmentalist and the evolutionist by regarding the first, like the physiologist, as asking *how* and the second, like the ethologist, as asking *why*-type questions. The evolutionary explanation combines the historical – though now with regard to an entire species rather than one individual – with the goal-directed. Perhaps because of this, some sociobiologists argue that it is *the* fundamentally causal question, and dismiss other causal claims as merely 'functional'.

The fifth biologist, a molecular biologist, smiles sweetly. 'You have all missed the point. The frog jumps *because* of the biochemical properties of its muscles. The muscles are composed largely of two interdigitated filamentous proteins, called actin and myosin, and they contract because the protein filaments slide past one another. This behaviour of the actin and myosin is dependent on the amino acid composition of the two proteins, and hence on chemical properties, and hence on physical properties.' This is a *reductionist* programme, and is the way in which biochemists seek to describe living phenomena.

But note again that this is not a causal chain in the sense in which the physiologist uses the phrase. It is not a question of *first* one thing happening (the actin and myosin sliding across each other), *then* another (the contraction). If the word 'cause' is used at all here, it must mean something quite different from how it is used in physiology. The confusion about the several ways in which 'cause' is used has bedevilled scientific thinking since the days of Aristotle. Perhaps we would see things more clearly if we restricted our use of the word to clear temporal sequences in which first one and then another event occurs. Each of these events – the image on the frog's retina, the processing in the brain, the transmission down a motor nerve and the muscle contraction itself – can be *translated* into the language of biochemistry. And of course it is possible to describe this biochemical sequence in temporal terms too, in which one set of biochemical processes (the molecular events in the nerve), produces another (the sliding actin and myosin filaments) (see p. 89). At issue, then, is the relationship between the two temporal sequences, that of the physiologist and that of the biochemist. In later chapters I shall explain why I use the term 'translation' to describe how the description of the phenomenon of muscle contraction in the language of (at the level of) physiology may be replaced by a series of presumed identity statements in the languages of biochemistry, chemistry, and so on.

IT ALL DEPENDS . . .

Biologists need all these five types of explanation – and probably others besides. There is no one correct type; it all depends on our purposes in asking the question about the jumping frog in the first case. Indeed, it turns out that 'it all depends' is a major feature both of living processes and of biologists' attempts to explain them. The reason for asking the question will determine the most useful type of answer. It is in the nature of biological thinking that all types of answer are – or ought to be – part of how we try to understand the world. Biology requires this sort of epistemological pluralism – to dignify our fuzzy way of thinking with a more formal philosophical imprimatur. To focus on any subset of explanations is to provide only a partial story; to try to understand completely even the simplest of living processes requires that we work with all five types simultaneously. None the less, the way in which the sciences of biology have developed means that excessive deference is paid to the more reductionist type of explanation, as if it were in some way more fundamental, more 'really' scientific, or as if at some future time it will even make the others redundant. Biochemists and molecular biologists, and indeed the fund-givers who support our research – government, industry, charities – are trained to think and argue in this way. It has become not second, but first nature for us.[9]

BIOLOGY IN TIME

The concept of time, and the idea of a direction of 'time's arrow', are central to biology. For many if not most aspects of the phenomena which physics studies, 'time's arrow' is reversible: processes can be driven in both forward and backward directions. The properties of matter and the 'laws' defining interactions are generally assumed to be uniform in space and time, even though our own human understanding of those laws is itself historically determined. Time, history, becomes relevant to physics and chemistry only in the context of cosmology. For much of biology such simplicity does not apply. Although the properties of living systems and processes are of course entirely in accord with the principles of physics and chemistry, a full understanding of them lies beyond the regularities that characterize those sciences' objects of study. Living processes are complex, often irreproducible because historically contingent, and are hence also practically irreversible. The arrow of time runs in one direction only: the direction studied by the developmental and evolutionary biologists in the frog fable.

For biologists, humans are not the product of special creation by an all-wise and all-powerful deity, but the more or less accidental product of evolutionary forces working over almost unimaginable aeons of time. Evolutionary biology has to write a history of life that has persisted for some four billion years. Most of us (scientists as well, in our day-to-day lives away from lab and computer) find it hard to think beyond a few generations: our own, our parents' and our children's lifetimes, a century or so, is about all we can manage. Yet the time-scale about which we have to think is surpassed only by that of the cosmologists with their universe of times and distances measured in billions of years and millions of light years – and light travels, we should not forget, at some 300,000 kilometres *a second*.

Evolution over time is a central biological theme; the past is the key to the present. Life as we now know it results from the combinations of chance and necessity that comprise evolutionary processes. Necessity, given by the physical and chemical properties of the universe; and chance, contingency, by the radical indeterminacy of living processes which it will be one of the purposes of this book to explore. That is, the indeterminacy is not merely a matter of ignorance, or lack of adequate technology; it is inherent in the nature of life itself. Indeed, the great population geneticist Theodosius Dobzhansky asserted that 'nothing in biology makes sense except in the light of evolution'. However, I wish to go several steps further. Nothing in biology makes sense except in the light of *history*, by which I mean simultaneously the history of life on Earth – evolution, Dobzhansky's concern – and the history of the individual organism – its development, from conception to death. But I have a third step to take as well. We cannot understand why biologists at the end of the twentieth century think as they – we – do about the nature of life and living processes without understanding the history of our own subject, biology. For us too, the past is the key to the present.

BIOLOGY IN SPACE

The second deep theme with which biologists are concerned is that of structure. The three dimensions of space must be added to the one of time. Organisms have forms which change but also persist throughout their life's trajectory, despite the fact that every molecule in their body has been replaced thousands of times over during their lifetime. How is form achieved and maintained? What are living organisms made of? How do their parts interact? These, as the fable of the jumping frog suggests, are above all the provinces of present-day biochemistry and molecular biology. Perhaps because these parts of biology developed historically later than chemistry and physics, the reductive methods of analysis and forms of explanation that characterize biochemistry and molecular biology and with which we feel most comfortable have been those derived from and are most congenial to these more senior sciences. Physics and chemistry, as essentially analytical disciplines, aim to disassemble the universe into its component parts, determine their composition and identify the 'laws' (preferably given mathematical expression) that govern their interactions. This has meant that, following in their footsteps, much of biology has hitherto been essentially analytical, happiest when taking things apart, reducing them to their components and deducing the workings of the whole from the functionings of these fragments. Yet cells, organisms, are more than simple lists of chemicals. Their three-dimensional structures, still less their lifelines, cannot simply be read off from the one-dimensional strand of DNA. Today the task of a biology of structure has become to understand how to reassemble the components, to explain both form and its transformation and persistence through time.

HOMEODYNAMICS

One of the dominating motifs in biological thinking was provided by the physiologist Claude Bernard in Paris in the 1850s. Bernard, who among many other discoveries carried out some of the earliest systematic studies on what were later to become known as enzymes and hormones, saw living systems as explicable neither by vitalism (the belief that there existed some special 'life force' beyond the reach of chemistry or physics) nor by mechanism. He regarded stability as a major organizing physiological principle, and emphasized the constancy of what he described as the *milieu intérieur* – the 'internal environment' – of multicellular organisms, their tendency to work to regulate this environment in terms of temperature, acidity, ionic composition and so forth. This capacity he saw as providing a stable context in which the individual cells of the body can function with a minimum of disruptive turmoil. Seventy years later the American physiologist Walter Cannon generalized Bernard's concept by introducing the term *homeostasis*[10] – the tendency of a regulated system to maintain itself close to some fixed point, like the temperature of a room controlled by a central heating system and a thermostat. No modern textbook account of physiological or psychological mechanisms fails to locate itself within this homeostatic metaphor. But the metaphor of homeostasis constrains our view of living systems. Lifelines are not purely homeostatic: they have a beginning at conception, and an end at death. Organisms, and indeed ecosystems, develop, mature and age. The set points of homeostatic theory are not themselves constant during this trajectory but change over time. The organism switches its own thermostat. Organisms are active players in their own fate, not simply the playthings of the gods, nature or the inevitable workings-out of replicator-driven natural selection. To understand lifelines, therefore, we need to replace homeostasis with a richer concept, that of *homeodynamics*.

AUTOPOIESIS

To summarize: to put the organism and its lifeline back at the core of biology, to counter the gene's-eye view of the world that has come to dominate much current popular and even technical philosophical writing on biology over the past two decades, means replacing the static, reductive, DNA-centred view of living systems that currently pervades biological thinking with an emphasis on the dynamics of life. We need instead to be concerned with process, with the paradox of development by which any organism has simultaneously to *be* and to *become*, as when a newborn infant must be capable of sucking at the breast while at the same time developing the competence to chew and digest solid food, and with the continuous interchange between organisms and their environments. These processes of development transcend the crude dichotomies of nature and nurture, gene and environment, determinism and freedom. Instead we must speak of the dialectic of specificity and plasticity during development,[11] the dialectic through which the living organism constructs itself. The central property of all life is the capacity and necessity to build, maintain and preserve itself, a process known as *autopoiesis*.[12] This is why it is in the very nature of life and living processes themselves that we, as living organisms and specifically as humans, are free agents. Not free in the Sartreian sense of the first epigraph at the head of this chapter, but free in the older, Marxist sense of the freedom of necessity. We humans, more than any other life form on Earth, make our own history.

BIOLOGY AS HISTORY

How biologists interpret the world is not itself unproblematic, despite the emphases that I, just as much as those of whom I am critical, choose to put on my certainties about 'how things are'. The biological story I am telling – and its critique of other stories – is not some timeless and universalistic one. It is told, like all stories, from a viewpoint, a perspective shaped by my own background as a particular sort of biologist, a biochemist with a major interest in how the brain works. And it is constructed at a particular time in the development of the biological sciences, a time of huge and rapid changes in technique and accretions of facts and observations about the living world at all levels from the molecular to the global.

Just as individuals and species carry the weight of history on their shoulders, so too do the sciences. Biology – not the phenomena of life, but their scientific study – is itself historically constructed. The very fact that it developed in the shadow of physics, with physics' goals of mathematical rigour and idealized predictive capacity, has deeply influenced biological thinking today. One consequence has been the power of technological metaphor in biology, whereby living systems become analogized to machines (hearts as pumps, colons and bladders as sewage systems, brains as computers, immune systems as military organizations . . .) – thus reversing a much older tradition in many cultures in which the physical world too was regarded as if it were alive. It is a fun thought experiment to consider what might have happened had this tradition been maintained, and biology had developed as a modern science before physics did. Would we have tried to construct machines along biological principles and endeavoured to explain their properties by invoking biological analogies – transport systems perhaps depending on legs and joints rather than wheels, rather as early attempts at flying-machines mimicked the action of birds' wings? Such attempts failed, for good structural reasons, and technologies based on biological principles have been successful only in the last few years, with the advent of parallel distributed computer architectures based directly on analogies with the organization of the brain.[13] Such historical considerations should help us to avoid a simplified view of late-twentieth-century biology as a tale of straightforward triumph in which the dark, error-ridden past is conquered with the help of the bright shining light of truth.

So I begin by asking how we know what we know: what is the philosophical and social foundation upon which science – and biology in particular – can claim to base its 'truths' about the world we study? How much do today's favoured biological explanations depend on the prevailing historical social and ideological climate, and how much on the availability of particular technologies (microscope, ultracentrifuge, radioisotopes)? All science depends on an interaction between observation, experiment and theory. How do we observe in biology? What constitutes an experiment? How far are our observations and experiments constrained by our theoretical mind-set? Can we begin to think outside and beyond our own historical frame, and make the leap towards a more integrative biology? And, above all, what will such an understanding mean for our vision of ourselves as humans and our relationship to the myriad other living forms with which we share our planet?

NOTES

1. I am indebted to Mary Midgley's *The Ethical Primate* for this quotation.
2. Like so many other aspects of the arguments in this book, this point has been brought home to me by Hilary Rose in her book *Love, Power and Knowledge*. As far as possible in this book I will talk about life, living processes and living systems, and restrict the use of the word biology to its proper limits, the *study* of these processes and systems.
3. More recently repackaged as 'evolutionary psychology'. See the forthcoming Novartis symposium on reductionism, edited by Jamie Goode.
4. Steven Weinberg, *Dreams of a Final Theory*.
5. Ernst Mayr, *Towards a New Philosophy of Biology*.
6. An aside by Watson during his debate with me at the Cheltenham Book Festival, 1994.
7. For one conspicuous example see e.g. Lewis Wolpert, *The Triumph of the Embryo*.
8. Goal-directed explanations used to be called *teleological* in that they implied an almost conscious sense of purpose. The more modern term *teleonomic* was introduced by the evolutionary biologist Ernst Mayr in an effort to make such explanations more philosophically respectable, on the grounds that goal-direction of this sort could be given an evolutionary justification without having to invoke conscious purpose.
9. The philosopher of science Thomas Nagel has claimed that only reductionism *explains*; everything else merely *describes*.
10. Walter Cannon, *The Wisdom of the Body*.
11. Susan Oyama, *The Ontogeny of Information*.
12. This term was first introduced by Humberto Maturana and Francisco Vanela in the 1970s, and I am greatly indebted to their approach for my treatment here. For a fuller treatment see Maturana and Vanela *Autopoiesis and Cognition* and *The tree of knowledge*.
13. Steven Rose, *The Making of Memory*.

2

Observing and Intervening

Life following life through creatures you dissect,
You lose it in the moment you detect.
Alexander Pope, *Moral Essays*, Epistle I

DOING SCIENCE

Doing science, finding out how the world works, seems obvious, unproblematic. We observe, collect facts, intervene, experiment, make hypotheses and test them, design powerful instruments to act as can-openers for those bits of the world not amenable to manipulation by unaided human capability. We publish our findings, and others use them to build further scientific knowledge, or to design technologies which profoundly alter the way we live. Doing this may be difficult, hard work, requiring heroic effort or inspired genius or collaborative, multidisciplinary teams. But there is surely no arguing with the method or the results it produces. In these technoscientific decades, everyone, except for a few troubled sociologists, philosophers, fundamentalists or romantic New Agers, takes it for granted. Such nay-sayers may stand outside the world's tent micturating in, but we have constructed it out of good man-made waterproof fabric, and inside it is warm and dry. Or is this not just a trifle smug?

Most 'working scientists', as we like to call ourselves (even though by the time we reach my present age we are mainly managers of others' laboratory activities, raising grants, writing papers and attending conferences rather than struggling with the intractable problems of making experiments 'work'[1]), are not troubled by – indeed are scarcely aware of – all this philosophical babble of metatheory. Our job is to get on with our trade, in physics, chemistry, biology or whatever, and try to tell it like it really is – to get to 'the truth' about the world. So in this chapter I want to ask how it is that we know what we think we know about the world – or more modestly, the world of living organisms and processes.

OBSERVING

It all begins with observation, with looking at the world around us. Observation is easy, obvious. Isn't it? Well, it all depends. I'm at a party in a crowded room, trying to hold a conversation face to face with someone I've just been introduced to. I ignore the babel of voices all around us, straining to hear what my new acquaintance is saying. Suddenly, from across the room, among all the sounds I have been shutting out, I hear my own name being spoken, and whirl around to try to find where the voice has come from. Psychologists call this the 'cocktail party effect'. We are constantly being bombarded with sensory stimulation from the world around us: sounds, sights, feels, smells. Most of this bombardment, for most of the time, doesn't get past our perceptual filters. And we even ignore most of the small fraction that does. Yet the fact that one can respond to the sound of one's own name spoken unexpectedly in the midst of the hubbub says that there must be some continuous monitoring process going on in the brain, observing and classifying the incoming data beneath the level of conscious awareness.

It is October, and I am walking in the beech woods with a Russian friend. Idly, I scan the tans, golds and purples of the fallen leaves. My friend Kostya is also focused on the leaves, but less idly. He darts forward, bends down and plucks from the variegated brown background an equally brown and to me hitherto invisible mushroom – a perfect boletus. In fact, until I met Kostya I wouldn't have known that it was edible, or indeed what to look for. Edible fungi hunting, pursued avidly by Russians, is a relatively rare sport in England. But once he has pointed out the boletus to me I soon match him as a hunter, spotting my prey where previously I would have seen only the myriad fallen leaves.

No one, not even a newborn baby, observes neutrally, innocent of preconceptions about the world outside. The baby scans and seeks a nipple and with experience rapidly learns to improve the primitive searching style already wired into its nervous system as a reflex at birth; an adult party-goer or mushroom-hunter hears their name called against a wall of sound, or finds the boletus against the almost identical background of the leaves. We continue, throughout our lives, to learn how to observe and what to select, what to define as object or foreground and what is field or background.

For years this interplay of observation and experience has been a hunting ground for perceptual psychologists. They have long played with images which their subjects are required to distinguish from their backgrounds, with ambiguous drawings which oscillate between alternative interpretations, and seemingly feasible objects which on close inspection turn out to be impossible (Figure 2.1). The fascination that our perceptions of such paradoxical figures have for psychologists

Figure 2.1 Ambiguous figures:
(a) Random dots or spotted dog? (b) Faces or vase?

lies in the conclusions that those perceptions make possible about the extent to which the world we observe is constrained – some would say constructed – by the architecture of our brains/minds. Autopoiesis – self-construction – is a major organizing principle of living systems. The issue of construction versus observation also lies at the heart of the paradox of science: that it claims to be able to provide us with something approximating to a 'true' account of the material world, yet it can do so only while viewing that world through prisms provided by the experience and expectations of its practitioners. This paradox has provided useful recent employment for philosophers and sociologists of science, some of whose work I shall turn to later in this chapter, but for now I want to pursue the question of observation a little further.

Science begins with systematic observation, an attempt to find regularities in the world around us, to predict future events on the basis of past experience. Suppose I am interested in how animals behave, and how that behaviour changes as they grow from infancy to adulthood. I may be watching a family group of marmosets, a pride of lions, a nest of hatching blue tits and their parents. I want to record how any of these spend their time during a day, a week, a month, a year. But I can't watch them continuously over the entire period, even if I were to mount video cameras and record every aspect of their activity. There would simply be too much data to analyse. That great Argentinian writer Jorge Luis Borges understood the problem well. One of his short stories centres on a character, Funes, with a total memory for everything that happens in his past. The problem is that he cannot forget, and the events of any day take the whole of the next day to recall; and Borges revels, as was his wont, in the logical paradox that this implies.[2]

So my first decision is that I must *sample* the behaviour. But for how long? Five minutes in every hour, one hour a day, one day a week? Do I watch all the animals in the group I am studying, or try to focus on just one? My decisions will depend partly on what questions I am asking about the behaviour, and partly on the resources – time, recording and computing power, or whatever – at my disposal. Perhaps I decide to make a video recording of the behaviour of the newborn twins in a family of marmosets, parents and their infants, by sampling ten minutes of their behaviour three times a day during the first few weeks after the babies are born. The twins interact with each other, with their parents, they suckle and cling, they begin to spend longer periods away from their parents. My video records patterns of continuously changing activity. But to make sense of it I need to classify, to distinguish the different types of activity I observe. In every ten-minute sample, how much time does any infant spend suckling, asleep, on the parent, off the parent, exploring, rolling on the ground with its twin . . . ?

Such a classification of behaviours and their distribution in time is called an *ethogram*, and to construct one requires active work on the part of the observer. It is necessary to decide which are the important distinctions to make between different aspects of the continual record of behaviour. Is scratching important, or is it only interesting when one animal scratches or grooms another? Which of the interactions between the twins counts as play – or is this not a meaningful category at all? If, during the first weeks of life, there is an increase in the proportion of the time recorded spent playing, is this a 'real' change, or is it an artefact resulting from the fact that the amount of time the infant spends asleep decreases, so the other recorded activities simply expand to fill up the sample time? The problem of separating object from field, of determining which is the 'correct' interpretation of the ambiguous figure, is not confined to the psychologist's abstractions but is the everyday stuff from which science has to be built.

Above all, which is object and which is field depends on the question one is asking. Ethologists often refer to what they call the 'four whys': questions originally posed by one of the founders of their discipline, Niko Tinbergen. Consider the question 'Why do birds sit on eggs?' The type of answer you want depends on where you put the emphasis in the question.

Why do birds sit on *eggs*? – that is, how do they recognize that eggs differ from stones?

Why do birds *sit* on eggs? – that is, why do they respond to an egg in this rather than in any other way?

Why do *birds* sit on eggs? – that is, birds as opposed to, say, mammals.

And finally, *Why* do birds sit on eggs? – that is, what is the function of this type of behaviour for the bird?

Until one is sure which of these possible questions one is asking, no meaningful observation and no scientific inference is possible. So, underlying any observation that we make of the world, even the most trivial – hearing a word, seeing a brownish mass on the ground – are the questions we wish to have answered (Is that my name I heard spoken? Is this an edible mushroom?). And inevitably behind these questions lie other questions – *metaquestions*. Why do we want to know the answer? What are the criteria by which we are prepared to agree that the question has been answered properly? And what type of answer would we find satisfying? The fact that within the theoretical framework that surrounds our observation or experiment we take these metaquestions for granted does not mean that they are straightforward, or that they do not present quite profound problems. Yet they underpin everything we do. Most of us spend our days living and working in buildings, of whose foundations we remain contentedly unaware, even though these are – literally – fundamental. Improperly constructed, the edifice collapses.

INTERVENING

So far I have considered only the question of observing events and processes going on in the world, without attempting to intervene in them. But most modern natural sciences – other than, perhaps, cosmology – are about more than just passively observing and recording. They attempt to understand the world by actively intervening in it, by first *controlling* it and then *experimenting* on it. There are several reasons for this. The first is that the sheer dynamic complexity of the world makes it hard to understand. Everything is moving, in constant flux, and unpredictable events disturb the regular pattern of our observations. The marmoset family is moving out of the focus of the video recorder, or reacting dramatically to the unusual sight of a snake just at the time we want to sample the family's routine activities. Our tidy ethogram is about to be upset. To extract meaning, we have to simplify, to try to keep the family in the picture and the snake out of it. Perhaps this means confining the group in a cage, regulating temperature, systematizing day length, providing food at regular times, and so on.

But also, as we begin to make predictions on the basis of our observations, it becomes necessary to test them. We can wait until some spontaneous event provides a 'natural' test, for the snake to appear of its own accord, or we can intervene to place it there at a time and manner of our choosing. This turns observing into experimenting. How much of the development of the twins' behaviour depends on their mutual interactions? Test by removing one of the twins and rearing it separately by hand. How much does the sex of the infants matter – do males play differently with each other than the way two females or a mixed-sex pair would? Test by swapping the partners. How much does the interaction of the mother with the infant depend on some characteristic odour – a secreted chemical pheromone? Test by filling the atmosphere with some novel odour – geranium oil, for instance; or try temporarily blocking the nostrils of mother or infant with wax.

Experimenting demands that we first simplify and control the phenomena we are trying to understand, and then intervene in them by changing variables systematically, holding all other things constant. The secret of the success of modern science lies in the development of this experimental, interventionist method, generally assumed to have been invented in the seventeenth century and given theoretical justification in the writings of Francis Bacon.[3]

The Baconian strategy is inherently interventionist. Indeed, its overenthusiastic application is said to have resulted in Bacon's own death from a chill contracted after he got out of his carriage in the depths of winter to carry out an experiment with snow as a means of preserving meat. It is also inherently reductive, because it works by attempting to isolate from the flux of the everyday world just the one aspect, the phenomenon, that we wish to study, and then changing one at a time the conditions we believe may affect it. If we change two or more variables simultaneously, we can never be certain which is primarily responsible for any effect we observe as a consequence. So it is necessary to decide which features of the experimental situation we are going to vary and which we are going to hold constant. If we think that, for example, the size of the pen in which we have confined the marmosets is important, we need to vary it; if not, it becomes part of the constant frame within which another feature can be modified.

But of course in the real world outside the confines of the laboratory lots of things are changing simultaneously. Variables and parameters become less easy to separate out. Effective experiments demand the artificial controls imposed by the reductive methodology of the experimenter, but we must never forget that as a consequence they provide at best only a very simplified model, perhaps even a false one, of what happens in the blooming, buzzing, interactive confusion of life at large, where things rather rarely happen one at a time and snakes intervene inconveniently.

The textbook example of this comes from precisely the type of controlled environment that I described a couple of paragraphs back – taking a monkey colony out of the wild and confining it in an enclosure in order to observe its interactions better. Back in the late 1920s, the anatomist Solly Zuckerman reported strong dominance hierarchies and high levels of 'aggression' and fighting among the large but confined hamadryas baboon colony at London Zoo, and developed an influential theory of social behaviour based on these studies. Each baboon, he wrote, 'seemed to live in perpetual fear lest another animal stronger than itself will inhibit its activities'. Violence was a constantly occurring event, quarrelling frequent and widespread, and any major disturbance of the precarious equilibrium caused the social order to collapse into 'an anarchic mob, capable of orgies of wholesale carnage'.[4] Later researchers observing baboon colonies in much larger enclosures or in the wild failed to find similar levels of fighting. Instead, the groups seemed relatively peaceful and stable. It became obvious – and with hindsight it seems scarcely surprising – that the behaviour of Zuckerman's baboon group had been dramatically modified by restricting the space within which its members had to coexist.

The constraints of Zuckerman's reductive approach had transformed the situation he wished to study and fundamentally misled him, even though his observations within that constrained situation were presumably perfectly accurate. Reductive methodology has served the simpler sciences of physics and chemistry well for three hundred years, and it is still the method of choice for most of the experimental work biologists do – including my own. But it may be failing us in our attempts to solve the more complex problems presented by the living world with which the biological sciences must now wrestle.

Take my own research as an example of a reductive, interventionist strategy at work. I am interested in memory – or at least in what happens in the brain when memories are made. My experimental 'model' is the young chick. I take pairs of chicks, put them into small, high-sided aluminium pens, 20 by 25 centimetres in area, and offer them little bright beads to peck at. They will usually peck at the bead within a few seconds. Some do so once, while some peck repeatedly. Some grab hold of the bead with their beaks and let go only reluctantly. Some peck sharply, seemingly angrily. One or two back away with shrill cries of distress. A few may be busy with other things – dozing gently, pecking at their companion's eyes or at the side of the pen, or preening their wing feathers – and refuse to be distracted. From this variety of behaviours I pick just one item to examine and record: whether, once the chick has clearly seen and paid attention to the bright bead a couple of centimetres in front of it, it will peck at it within the twenty seconds I allow. It is this much-reduced region of an ethogram that I take as my starting-point.

Such 'pecking responses' are among the most basic of possible 'observations' to make for someone interested in animal behaviour. Yet notice what I have done in order to make these seemingly straightforward observations. First, I have reduced my field of study by arbitrarily simplifying its context, confining the chick in a more or less empty environment, devoid of any significant cues which might interest or distract it. Then I have introduced a specific complexity into the environment by putting a second chick in with the first one. My reason for doing so is that chicks prefer to be with their companions: they show fewer signs of distress, and hence are more likely to pay attention to the bead than if they are isolated. So to avoid one problem in my experimental design I have introduced another potential confuser. Not only have I picked the two companions at random, but I choose to ignore any of the subsequent interactions between them, as I am interested in only one thing – whether or not the chick I am studying pecks the bead in the time I allow it. This constrained, highly artificial situation will form the terrain for any subsequent experimental intervention I choose to make – and I am immediately in danger of falling into Zuckerman's trap. If I am to avoid it, I must be very clear indeed about the type of data I wish to extract from my experiment.

But I am already in another sort of trouble. Think for a moment about the words I have used on the previous page to describe the different ways in which the chick approaches and pecks at the bead: 'reluctantly', 'angrily', 'distress'. All these words seem to have clear-cut meanings when we use them to describe how we ourselves behave. But by what right can they be applied to a chick's behaviour? The chick can't tell me whether it 'really' feels like that; over the many years in which I have observed chicks, I arrogate to myself the right to categorize, and indeed to anthropomorphize, their behaviour in this way. But whatever the word I might use to describe how the chick pecks at the bead, in making my observation about that feature of the behaviour which interests me, I have ignored everything except the one measure: did the chick peck? Because I can't quantify or measure 'objectively' the various *kinds* of peck that the chick may make, all I can do is answer the simple yes/no question of whether a peck has occurred within a given time.

My observation may thus be objective, in the sense that anyone else observing the chick at the same time as me will record the same event, and indeed it could be captured by an automatic recording device with no human interference at all – all I would have to do is put inside each bead a little sensor that responds to the pressure of the peck. Yet even then my objective observation and measurement is at the same time subjective, for it does demand a sort of skill on my part. Odd though it may appear, it turns out that not everyone can train chicks to perform this simple task. I have occasionally had students in the lab for whom the birds simply won't perform well – they find disturbing something about the way a student approaches them. And there is subjectivity in a more fundamental sense too. In designing my experiment, I have chosen to observe and record what seems to me to be important in relation to my real interest: memory.

How so? Consider the next part of my experiment. This time, instead of simply offering the chick a dry bead to peck, I first dip the bead in a bitter-tasting liquid. The chick will peck it once, shake its head vigorously, wipe its beak on the floor of the pen – and then refuse to peck at a similar but dry bead offered anything up to several days later. This is the crucial experimental intervention I am making into the life and activities of the chick, and it is the basis for everything that follows, for I attribute this refusal to peck at the bead to the chick having a 'memory' of the bitter taste of beads of this particular size, shape and colour – indeed, I *define* the refusal in these terms.[5]

METAPHORS, ANALOGIES AND HOMOLOGIES

By defining an observation about an activity on the part of an animal as a particular exemplar of a general phenomenon, 'memory' (a bitter memory in this case!), I have given myself a lever with which to move at least part of the world. I can then go on to ask what happens in the brain of the chick when it 'learns' that the bead tastes bitter, or when it 'remembers' the taste when later offered a similar bead to peck. And by subsuming these very specific bits of behaviour on the part of the chick within such categories as learning, remembering and memory, I imply that the processes I am studying are in some way related to those which we also call learning, remembering and memory in frogs or snakes, rats or snails – or humans.

Note also, though, that such methods, which are used to investigate these internal processes within the chick's brain, are inherently not merely interventionist, but violently so. I have had to kill the chick to observe the changes in its brain. My subject of study is at the same time the object of my (terminal) intervention. This is of course one of the paradoxes of the reductive methodology in biology. It is a paradox we may deplore, but we cannot avoid it if we are committed to the belief that the information we shall acquire, the theories we may build, from such a process can tell us something of value about the world. How we define 'value' distinguishes the interventionist approach of this type of biology from the cruelty of badger-baiting or the idle curiosity of pulling the wings off flies. It demands moral judgements, made by the researcher and also by the society that sanctions the research. In my case, if I am right, what I discover about the cellular processes of memory in *Gallus domesticus* will apply also to *Homo sapiens*, opening up the prospect not merely of knowledge but also of potential therapeutic intervention for sufferers from the losses of memory suffered by those who contract brain diseases such as Alzheimer's. I shall be able to make a generalized statement about memory, based on the following logical sequence:

1. Chicks which avoid pecking a bright bead after having pecked it once and found it bitter are showing memory for the association between the appearance of the bead and its taste.
2. This behaviour is reflected in certain necessary and specific changes in the brain of the chick, and I can study those changes.
3. Human brains resemble chick brains in certain fundamental ways.
4. Therefore, when humans show memory, similar changes are going on in their brains.
5. And therefore what I learn about how to intervene in the processes of chick memory can be applied to human memory too.

The validity of this syllogism depends crucially on the third of these propositions.[6] There are three ways in which 'resemblance' can occur, and everything revolves around which of the three applies in this case. Is the process I am studying in chicks best regarded as a metaphor for human memory, analogous to it or homologous to it? Biology uses all three terms, but they are quite distinct in meaning and significance.

In a *metaphor* we liken some process or phenomenon observed in one domain to a seemingly parallel process or phenomenon in a quite different domain. For instance, during the 1930s an almost universal biochemical process was discovered whereby the energy released during the oxidation of glucose and other foodstuffs was trapped and held in available form within the cell by using it to synthesize another molecule, called ATP (adenosine triphosphate). ATP became described as the 'energy currency' of the cell, and the storage and flow of energy via ATP and related molecules was likened by one of its discoverers, Albert Lehninger, to the workings of a bank. ATP was the cell's current account; other molecules (such as creatine phosphate) served as the 'deposit account'. Wages, in the form of glucose, were paid into the account; work done by the cell, in synthesizing proteins or in muscular contraction or whatever, required withdrawals of ATP currency from the bank. The power of the social baggage that comes along with such a metaphor should not be underestimated, for it shapes the ways in which experiments and hypotheses are designed. More recently, the variety of metaphors for DNA and its genetic functions have grown almost out of hand – it has been likened to a codebook, a blueprint, a recipe and a telephone directory, to name but four of the more prosaic comparisons (I shall leave for later chapters the more grandiose references to the human genome as the 'Book of Life' and the 'Holy Grail'). Metaphors are not meant to imply identity of process or function, but rather they serve to cast an unexpected but helpful light on the phenomenon one is studying. None the less, as I shall argue in Chapter 6, their seductive charm is highly dangerous.

Like most such terms, analogy and homology have multiple meanings. In the context in which I am using them here, *analogy* implies a superficial resemblance between two phenomena, perhaps in terms of the function of a particular structure. Thus there are ways in which it makes sense to consider the blood circulating in animals as having functions similar to – analogous to – those of the sap in a plant; and as a more mechanical analogy, the heart can be regarded as a pump. Such analogies can be quite precise. After all, hearts can be replaced by artificial pumps, and the mathematics that describes the heart's action in driving blood through the circulatory system is the same as that used to describe the functioning of a water pump in a car engine.

But analogies can also mislead – is it a help or a hindrance to regard the random access memory (RAM) in my computer as analogous to memory in chicks or humans or is it merely a metaphor? Again, this issue will return to haunt later chapters.

By contrast, *homology* implies a deeper identity, derived from an assumed common evolutionary origin. This assumption of a shared history implies common mechanisms. It is in this sense that the bones of the front feet of a horse may be regarded as homologous to those in the human hand, and, I want to argue, that chick memory is homologous to human memory.

Is it legitimate to argue for homology between what happens when a chick pecks a bitter bead and what happens when you or I try to remember a telephone number? That is, do those aspects I have extracted from the continuous processes by which the chick interacts with its environment really represent some unitary feature of the material world, one that may be distinguished from any other? I claim that they do, but my right so to claim is not self-evident. The issue goes beyond the fact that I have to argue convincingly that these two superficially very different-seeming activities are both exemplars of a more general phenomenon. There is a fundamental issue at stake here.

Can I extract, from the continuous process by which the chick or I experience and interact with our environment, a discrete entity called 'memory'? This raises a question which goes to the heart not just of the scientific method but of philosophical traditions running back many thousands of years. In general there are two ways of looking at what goes on in the world around us. In the more familiar, which derives from the cultural heritage of the Judaeo-Christian and Graeco-Roman traditions within which modern science is done, the world is composed of isolable entities – electrons, or atoms, or molecules, or organisms, or tables and chairs – which possess discrete properties, such as memory, and interact with one another according to definable laws. In the second, less familiar view, the world is one of continuous process, out of which transitory entities occasionally crystallize. We are dealing again with distinctions between object and surround, foreground and background. This latter way of conceptualizing the world is perhaps more akin to non-Western philosophical traditions, such as those of India and China. But for most of the past hundred years, theorists have had to come to terms with such a world-view, for instance when they alternate between treating light as a stream of particles and as a wave, or when their mathematical symbolism demands that they speak of magnetic or gravitational *fields*. As I shall argue, many of the problems in the biological sciences derive from the cultural difficulty we have in perceiving a world of fields and processes rather than of objects and properties.

NATURAL KINDS

The object-centred view of the world was given philosophical form by the Greeks. For Aristotle, the world is composed of observable phenomena, underlying which it is possible to define a particular essential set of properties, or *natural kinds*. On the surface, this view of the world (which Aristotle derived from his predecessor Plato) resembles anyone else's. It is full of objects: tables and chairs, cats and dogs. Each forms a separate category, even though each can take many different forms. Tables can be large or small, can be made of plastic or metal or wood, and can stand on one central leg or several peripheral ones. Dogs can be Saint Bernards or poodles or dachshunds. But below the surface, the Platonic world takes on a different reality: underlying all tables is the essential ideal table, under all dogs the essential ideal dog. The tasks of philosophy and of science are then to identify and define these essences underlying surface reality, and hence of dividing the world of things and processes into its 'natural' units, a procedure described as 'carving nature at its joints'.

Within the world of human artefacts it may not be unreasonable to try to seek the essence of, say, a table or chair. We can define them by purpose: a chair is to sit on, a table is to bear objects we wish to access while sitting on a chair. Numbers of legs, colour, even shape or size within limits, can all be modified without affecting these essential functions. Such a view of the world may even be possible when one is studying inanimate phenomena such as comets, electrons or chemical elements. I don't want to get into an argument over this;

astronomers, physicists and chemists must speak for themselves on such topics. My concern is with living systems.[7]

Are there 'natural kinds' and clearly defined joints at which to carve nature in the living world? At first sight the answer would seem obvious. Every reader of these words is an individual, a person, a member of the human species. So is there an essence of humanity which enables us to clearly define what constitutes a human individual? Most of us have no difficulty in recognizing adult, or even infant fellow-humans. So it would seem that the answer is that there is an essence, even something we can define as a 'universal human nature', masked though it may be by the preoccupation of some among biologists (see Chapter 7) to privilege differences over similarities.

But think of the difficulties faced by moral philosophers, Catholics or embryologists wrestling with the problem of defining where human life begins, at which point the fertilized ovum or embryo must be allowed to have those 'rights' which are (or at least should be) inalienable for humans. Or of defining death for a person on a life-support machine. Or, in terms of human evolutionary history, of reaching agreement on which of the various fossilized human ancestors that have been discovered over the past century can properly be regarded as human. The fact is that, whereas in pre-Darwinian days species were regarded as immutable natural kinds, each the product of divine creation and for ever distinct, modern biology has great difficulty with the concept of species and its boundaries in space and time. Even the most straightforward definition, as constituting a group of individuals capable of fertile mating, is thrown into disarray by some of the advances in current gene technology, which offers bizarre crosses such as the shoat, the engineered offspring of sheep and goats.

If we cannot take a species as an example of a natural kind, how about subdivisions within the category? Much of the history of two centuries of Western anthropology has been about the attempt to identify 'scientifically' acceptable divisions among human populations, divisions which could be classified as 'races'. In pre-scientific days there seemed little trouble with the term; English literature, social and political writing is full of references to the Scottish, Irish or Welsh race, in which the term 'race' is seen as encompassing certain cultural and historical inheritances which shape an individual's psychology and personality. In this typological way of thinking, races appear as Platonic natural kinds. Nineteenth-century evolutionary theory and anthropology seemed to offer a more biological basis for such distinctions. Races could be categorized on the basis of skin colour, or on skeletal or skull structure, and could even be arranged in a hierarchical order from the 'more' to the 'less' evolved.[8]

The sorry history of this scientific racism, a history made possible only through the enthusiastic collaboration of many psychologists, geneticists and anthropologists, has been told many times,[9] and there is no need to retell it here. As will become clearer in later chapters, modern population genetics makes the concept of 'race' in the human context biologically meaningless, although still socially explosive.[10] The definition of race is essentially a social one, as in a reference to Blacks or Jews. While there are differences in gene frequencies (that is, differences in the proportions in which particular genetic variants occur) between population groups, these do not map onto the social criteria used to define race.[11] For instance, Polish Jews resemble genetically their fellow Polish nationals, non-Jews, more closely than they do Jews from Spain. Gene frequencies in Black Americans differ from those in Black South Africans. And for that matter, gene frequencies differ between people in North and South Wales, yet no one would think of classifying those two populations as two different races. This typological thinking has not disappeared: it characterizes, of course, the poisonous propaganda of racist political groups, and has not entirely vanished from popular scientific writing.[12]

If species and races have at best fuzzy boundaries and at worst are empty categories, how about individual organisms? Aren't I, writing this chapter, and you, reading it, units with clear boundaries? But where do our boundaries lie? We can cut our nails or hair without feeling that we are losing part of ourselves as individuals. We can contemplate having a limb amputated, or losing sight or hearing or the power of speech. Each of these disasters may in larger or smaller measure reduce us as individuals, but we still retain our sense of unity, however diminished – the amputated limb is no longer part of us. (Oliver Sacks has written fascinatingly about the distressing and bizarre consequences of certain sorts of brain damage as a result of which parts of one's body are no longer perceived as 'self' but rather as alien objects.[13])

And these days we practically take for granted having parts of the external world inserted into us: ceramic teeth, titanium hips, battery-driven hearts or pigs' kidneys – even other people's 'spare parts'. We assume that such insertions will in some sense come to be part of us, assimilated into our sense of our specific, unitary nature. Of course, these issues are not straightforward: consider the moral unease generated by the use of genetic engineering techniques whereby 'human' genes are inserted into mice or bacteria in order to generate either 'models' for human diseases or 'factories' which can synthesize commercially or clinically desirable products.

Even disregarding such intrusions or deletions, the borderline between the organism and its environment – the definition of what actually constitutes you or I as an entity – is still no simple matter. Most of us will have experienced at some time or another the intense joy of losing our sense of boundaries completely during sex. And for a pregnant woman the differentiation – or lack of it – between self and the foetus she carries is prolonged and complex. But even these examples apart, look more closely at any human body. We are built out of tissues arranged into organs, each tissue a mass of individual cells, each cell an assemblage of molecules. We may expect to live for seventy, eighty, ninety years or more. During that time every cell in our body (with the exception of the nerve cells – neurons – in our brain) will have died and been replaced many hundreds or thousands of times. And every one of the giant macromolecules – the proteins, nucleic acids and lipids – of which the cells (including neurons) are constructed will have been laboriously synthesized, and persisted for a few hours, days or months, only to be broken down again and replaced by a successor molecule, a more or less exact copy. Our bodies are in continuous flux. Nothing about us as organisms is permanent. Wherever our sense of unity and individuality in space and time comes from, it cannot be from the persistence of the molecules or cells that comprise our bodies. Our sense of self is generated for each of us through the identity provided by our lifeline. It does not derive simply from the persistence of molecules or cells, or even bodily structures, which are transient, but by our life processes, which continue dynamically throughout our existence. This is a *process* unity, rather than an *object* unity.[14] Once more, this is why we are defined as individuals by our history at least as much as by our molecular constituents.

Nor are our boundaries impermeable. Our guts harbour hundreds of millions of micro-organisms (predominantly the ubiquitous *E. coli* bacterium) living symbiotically or parasitically with us. Many more minuscule living creatures inhabit the surface of our body, skin and hair. Some we are conscious of, often unpleasantly so; others not. We don't normally regard them as contributing to our sense of individuality, yet deprived entirely of these other living forms which share our personal space with us, we would scarcely be able to survive. What may thus seem at one level of magnification, and for much of the time, a clear-cut division between any individual and the outside world which forms his or her environment ceases to be so the moment we look closer. Humans are more coherent than a colony of corals, but the definition of where we begin or end in either space or time is fuzzy, not sharp.

So, if not among species, races or organisms, where can we find natural kinds in living systems? How about down among the molecules? I referred above to the macromolecules of which our bodies are composed. Take proteins, for example: molecules constructed from linked chains of smaller sub-units, the amino acids, of which there are some twenty variants. Each protein consists of a unique sequence of several hundred amino acids. This sequence is known as the *primary structure* of the protein. But the chain is coiled up in helical and pleated patterns, wound back on itself into a configuration which is held into shape by complex arrays of electrochemical forces (these patterns and arrays are known as the *secondary* and *tertiary structures*). Within this globular mass are trapped other, smaller molecules and ions – hydrogen ions derived from water, and metals including calcium, magnesium and iron (Figure 2.2). Deprive the protein chain of these smaller ions or molecules, or shift the acidity or alkalinity of the solution in which it is dissolved too far from neutrality, and the globular structure

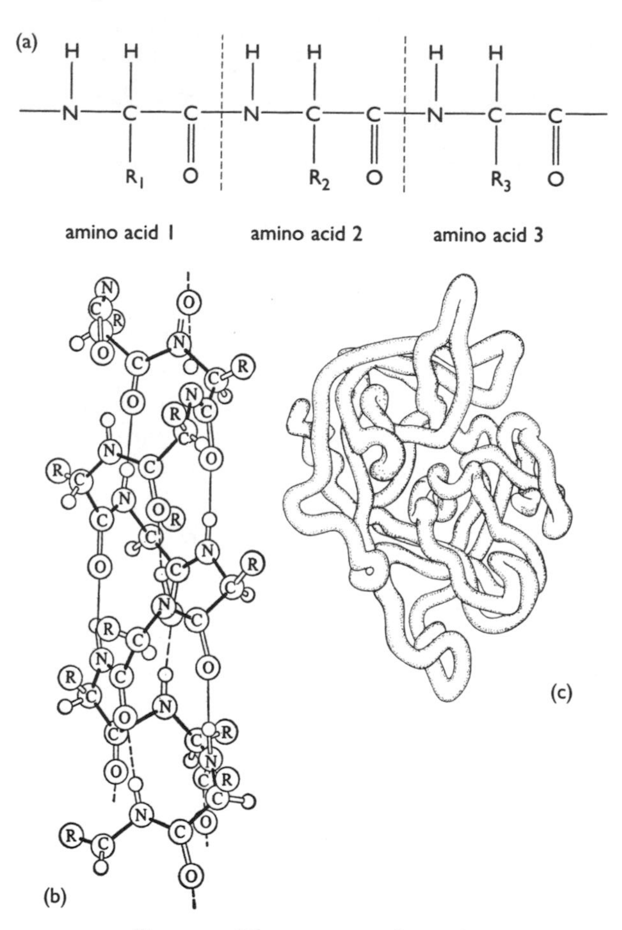

Figure 2.2 The structure of protein:
(a) primary, (b) secondary and (c) tertiary.
R is an abbreviation for the rest of the amino acid molecule.

collapses, often irreversibly – this is what happens when milk curdles, for instance. Furthermore, proteins in living cells do not exist in isolation. They are linked with other proteins into higher-order (*quaternary*) structures, or embedded in lipid membranes, or tightly bound to RNA or DNA. So how do we define the protein? By its primary sequence or its tertiary structure in space? Do we include all the ions and molecules it collects around its surface and within its crevices? What constitutes the Platonic essence of the protein – or is there no sensible way we can ask this question?

Perhaps we can distinguish the protein by its function rather than by its structure. Such a functional definition runs into other problems, though, for it turns out that organisms often contain several variant forms of the primary structure of any particular protein (*isoforms*), which appear to be functionally equivalent as far as the organism is concerned. Nor does it seem that all the amino acids in a protein chain are functionally necessary, for it is generally possible to lop off or add amino acids to the chain without apparently affecting the part the protein plays in the economy of the cell. However, some regions of the molecule are essential to its function, and are interfered with only at peril. For instance, the substitution of just one of the 146 amino acids in the β-chain of haemoglobin – a valine for a glutamate at one particular position in the chain – results in a change of properties of the molecule, the 'sickling' of the red blood cells in which the haemoglobin is contained, with a consequent risk to the life of the person who carries the variant form of the molecule. So a functional definition of any particular protein would produce only a partial overlap with a structural definition.

It seems that the more molecular biologists and biochemists discover about macromolecules, the less certain the picture becomes. For instance, one major class of proteins is the enzymes, molecules that serve as very specific chemical catalysts inside the cell and enable high-precision transformations of other molecules to be executed. It used to be believed that all enzymes were proteins, and older textbooks offer this protein nature as one of the defining characteristics of enzymes. A few years ago it was discovered that certain types of RNA molecule could also function as enzymes, and they were promptly christened *ribozymes*. The definition of what constitutes an enzyme could no longer be made on the basis of structure, and now rested on function alone.

Thus, while it is possible to offer a general definition of a protein as a molecule composed of a long chain of amino acids linked in a particular way, any decision as to whether a particular protein is defined according to primary sequence, tertiary structure or function, or about which of its ionic and molecular encrustations to include in the definition, can only be operational, depending on the purposes for which we need to make the definition. A protein is no more a clear-cut natural kind than is an organism or a species. And the same is true for the other macromolecules of which the cell is composed, polysaccharides and lipids. As we shall see, it is true even for that mythopoeic molecule DNA, nowadays regarded as first among macromolecular equals, having displaced proteins from their biochemical primacy – even though their very name derives from the Greek for 'first things'.

Thus, even though they give the superficial appearance of carving nature at the joints, definitions – 'essences' – in biology are always operational rather than absolute. Even at their best, they are fuzzy at the boundaries. At their worst, like the definitions of 'race', they may serve only to obfuscate, to pretend to differences that vanish or become unsustainable on closer inspection. 'Good' definitions are good because they are adequate for the purpose we intend them for, as they help us classify and order the world we observe. But we would be wrong to imagine that definitions have primacy over the observations upon which they are based, that they are in some way revealing a Platonic essence that exists prior to and independently of the observations which call them into existence and the purposes for which we wish to use them. In a world which is understood in terms of process rather than object, the joints into which we carve nature depend on our ultimate purposes, just as do those into which human carnivores may carve slices of roast meat for the table, or an artist a tree into a wooden sculpture. Certainly they have to bear some relationship with the material world: we cannot alter butchery styles entirely at will, observe phantoms, carve imaginary objects or force them into configurations entirely of our own volition. But we do have choices, and these choices depend on an interplay between the nature of the world we are studying, our understanding of what type of answer to the questions we will accept, and the reasons why we are asking them. In the next chapter I consider how and why we make such choices, and the extent of their validity.

NOTES

1. See Bruno Latour, *Science in Action*, for a dissection of the meaning of being a scientific manager.
2. Jorge Luis Borges, 'Funes the memorious'.
3. For discussion of Bacon, see Charles Webster, *The Great Instauration.*
4. Claire Russell and W. M. S. Russell, *Violence, Monkeys and Man*; the quotes are from p. 41, and the account of the later research is from p. 43 onwards. For his own account, see Solly Zuckerman, *The Social Life of Monkeys and Apes.*
5. I've written about how I use this model to study memory elsewhere (in *The Making of Memory*), and I don't want to tread that ground again; the purpose of the example here is different.
6. It is precisely this which one wing of the animal-rights movement disputes, claiming that what I argue is homology is at best metaphor.
7. My friend and colleague Brian Goodwin takes exception to this section of my argument. He points out that natural kinds do not have to be static unchanging entities. They can be defined by generative mechanisms, and their invariant properties are thus only relatively invariant – for a more detailed argument, see Gerry Webster and Brian Goodwin, *Form and Transformation.*
8. Carleton S. Coon, *The Origin of Races.*
9. See e.g. Stephen Jay Gould, *The Mismeasure of Man.*
10. That not all biologists understand this – even some, such as Steve Jones, who ought to know better – is a continuing source of public confusion on the matter; see, for example, the muddle Jones gets into on the matter in his coffee-table book *In the Blood*, and contrast it with Richard Lewontin's account *Human Diversity.*
11. See the discussion of the definitions of race by Rose, Lewontin and Kamin in *Not in Our Genes.*
12. For a discussion of the re-emergence of racial biology, see e.g. Marek Kohn, *The Race Gallery.*
13. Oliver Sacks, *The Man who Mistook His Wife for a Hat.*
14. Again, this is a normalizing statement. Unitary senses of self can be lost in diseases such as schizophrenia, or discarded by the psychotherapeutic interventions that seem to generate the type of learned behaviour described as multiple or dissociative personalities – see Ian Hacking, *Rewriting the Soul.*

3

Knowing What We Know

No scientist is admired for failing in the attempt to solve problems that lie beyond his competence. The most he can hope for is the kindly contempt earned by the Utopian politician. If politics is the art of the possible, research is surely the art of the soluble. Both are immensely practical-minded affairs.

Peter Medawar, *The Art of the Soluble*

INDUCING AND DEDUCING

The purpose of observing and experimenting is to derive knowledge of the material world and its workings; to enable us as individuals, and society as a whole, to understand, to predict and in some measure to control that world, to mould it to our purposes. This action imperative was there in modern science from its beginnings, and is far removed from the contemplative reflection on nature and fate which had characterized earlier forms of scholarship. Francis Bacon showed that he clearly understood the potential of the new science when he described experiments as being of two kinds: those that brought light, and those that brought fruit.

For Bacon, the way in which the experimental method provided reliable knowledge of the world was straightforward. You collected facts. You made an observation, or performed an operation on the world and noted the consequences. If the same observation or operation was repeatedly followed by the same consequences, you could reliably draw the conclusion that this indeed reflected the way the world was organized. If you throw a switch on the wall and the light in the room comes on, it may just be accident or coincidence. If you do the same a second time and the light comes again, you may well suspect that the two are causally related. A third, a fourth and a fifth time, and you may be pretty sure you are right. This is the Baconian method of *induction*, and for nigh on three hundred years after he formulated it, it was the way in which most scientists believed they worked. But it has a fatal flaw. No matter how often you throw the switch and the light comes on, you cannot be certain, in the absolute sense that philosophers demand, that the same thing will happen the next time you do it. The fact that – so far as I know – every human who has ever lived has eventually died, and that I am human, makes me pretty sure that I shall die too. But maybe I am wrong. Could I not just be the exception? Perhaps death is not an *inevitable* corollary of life.

Charles Darwin, who claimed that he was no philosopher, was none the less quite clear that he at least did not do science in this way. As the great observer, systematizer and collector pointed out, facts have no meaning in themselves until they are collected and presented *for* or *against* some hypothesis. The philosopher Karl Popper articulated this alternative view of science in a form that many found irresistible – at least for a time.[1] Science proceeds not by *in*duction, Popper argued, but by *de*duction. Scientists make hypotheses about how the world works, consider the implications of their hypotheses and design experiments to test them. A hypothesis might be that the light comes on whenever you throw the wall switch because the switch activates a beam of infrared radiation which triggers some sensor at the light bulb. However, no matter how often you verify that the light comes on whenever you throw the switch, this won't help you get to the truth of the matter. You could test the hypothesis by showing that the throw of the switch did indeed result in an infrared pulse, and that there was a sensor at the light bulb that was sensitive to it. But even this wouldn't prove that the hypothesis was correct. What you need to do is to design a crucial experiment – one which deliberately sets out to try to *falsify* the hypothesis – for instance by putting a heavy metal screen between the switch and the bulb to block out any possible radiation. If the bulb still comes on when you throw the switch, it can't be doing so by infrared radiation, as the screen would block it, so your hypothesis is shown to be false, and you need to make another one – perhaps that throwing the switch completes a circuit of hidden wires connected to the bulb. If, on the other hand, the bulb doesn't light up when the screen is in place, your hypothesis is strengthened and lives to fight another day. None the less, for Popper all hypotheses are provisional, good only for as long as they can withstand attempts to overthrow them. And the best hypotheses are those for which one can most readily design falsifying tests – crucial experiments.

Popper's thesis, originally formulated in the 1930s, was rapidly adopted by many philosophers of science, but it was only when his views were explained to us by one of our own – the immunologist Peter Medawar, whose words form the epigraph to this chapter – that most researchers realized that we had never really worked as Baconians. We were, above all, hypothesis-makers. So enthusiastically were Popper's ideas taken up that during the 1970s and 1980s grant applications to Research Councils in Britain tended to be turned down if they failed to state that the purpose of the proposed research was to 'test the hypothesis that . . .'. Mere Baconian fact-collection was no longer sufficient. (In the dourer 1990s, even hypothesis-testing is no longer the key criterion; instead we have to show 'relevance' to 'wealth creation'.[2]) Popper became to all intents and purposes the only philosopher of whom natural scientists in the anglophone world had ever heard. He was certainly the only one in modern times to have been made a Fellow of the Royal Society, and his death in 1993 resulted in an obituary and a stream of correspondence in *Nature*, science's journal of record. None the less, it is doubtful that many of *Nature*'s readers recognized the mortal blow that Popper had struck against our deeply held conviction that we were engaged in discovering the 'truth' – or at least 'truths' – about how the world works. After Popper, absolute truths no longer existed, merely provisional hypotheses, constantly under threat from new challenges.

There is a double irony here. In the first place, some of the central theories in science, notably that of evolution by natural selection, are by Popper's criterion unscientific because they are unfalsifiable. As must be the case for all essentially historical theories, it is not possible to design an experiment which (to use the Popperian term) disconfirms Darwin. To adapt a metaphor used by Stephen Jay Gould in his superb book *Wonderful Life*,[3] to test evolutionary theory in a Popperian manner you would need to wind the tape of history back and replay it time and again under a variety of different circumstances – an achievement possible only in certain model test-tube experiments. Yet no biologist would for a moment consider abandoning the theory on the mere say-so of a philosopher. Popper later modified his falsifiability criterion to take account of this problem, but by the time he did so hypothesis-making and falsifiability had become the chapter-and-verse taught to fledgling scientists in schools and universities, at least in Britain.

POPPER VERSUS PARADIGMS

The second and greater irony is that, just when natural scientists emerged into the bright Popperian light – when Popper had become part of the common sense of science, his model of how science proceeds came under attack from his own peers among philosophers, historians and sociologists of science. The first assault was led by the historian Thomas Kuhn. His argument, based on the history of physics, was that for most of the time scientists aren't doing anything as grand as making and testing hypotheses. We are simply solving puzzles set by the work of earlier researchers, within an overarching theory about the way our bit of the world works and which we are not concerned to challenge.

Kuhn called the work we do 'normal science' and our overarching theories 'paradigms'.[4] The example he gave was Newtonian physics. From time to time, research produces anomalous results – ones which cannot easily be accommodated within the accepted paradigm. The paradigm then has to be shored up with all sorts of supplementary hypotheses, so that it becomes more and more cumbersome. None the less, paradigms can always be saved – after all, the motions of the planets can be predicted quite well using the pre-Copernican, Ptolemaic system in which the Earth rather than the Sun is the centre of our immediate universe. Sooner or later, Kuhn argued, a new paradigm would emerge which transformed the world-view, shaking up all the old puzzles and setting them into a new framework. Thus the Copernican view of the world replaced the Ptolemaic in the seventeenth century, and became part of the Newtonian world-view which persisted until the beginning of the twentieth, when Einsteinian relativity offered a newer and more attractive paradigm with which to replace Newton. Kuhn called episodes in which one paradigm replaces another 'scientific revolutions'. His view was attractive to historians and sociologists of science, but it had a resonance for natural scientists too. Most of us who read Kuhn saw immediately that for much of our working lives we were doing work which was too humble to be called hypothesis-making or falsifying. We were, for most of the time, solving puzzles – doing 'normal science'. Few of us have the privilege of participating in a Kuhnian paradigm-breaking revolution.

Kuhn, like the majority of philosophers of science, took physics as his own 'paradigm case'. Biology offers fewer examples of either grand paradigms or paradigm-breaking experiments, presumably because we deal with much more varied and complex phenomena than are found in physics. Our paradigms tend to be rather smaller in scale, more local, less universalistic. There is no equivalent in biology to Newton's laws of motion. At least there seemed not to be until the 1990s, when efforts have been made to elevate so-called 'universal Darwinism' to a Kuhnian paradigm into which all phenomena of life must be shoehorned[5] ('shoehorning', by the way, is another metaphor which I have borrowed from Gould). A sub-paradigm within universal Darwinism is the DNA theory of the gene and replication. Thus, in the afterglow of Kuhn's book the historian of science Robert Olby retold what he called 'the path to the double helix' as an account of replacing a previous, protein-based theory of life with the new DNA-based paradigm.[6] I shall return to a critical discussion of both these paradigms in later chapters.

Breaking and remaking paradigms meets considerable resistance, for in many respects the scientific community – myself included – is rather conservative. Old paradigms, one might say, never die; only their protagonists fade from view. For instance, on a less grandiose scale than relativity or evolution, the belief that memories are stored in the brain in the form of changes in the properties of nerve cells and in the connections between them is the paradigm within which my own research is set, and within which I am largely content. Like other paradigms, it is hard to disprove. One alternative, energetically proposed by the botanist turned New Age philosopher Rupert Sheldrake, is what he calls 'morphic resonance'. His idea is that memories – human and non-human – are not stored in the brain at all, but are somehow present in a universal 'ether', so that once something has happened somewhere in the world, it becomes easier for it to happen again somewhere else.[7] His books and public appearances attracted a good deal of non-scientific enthusiasm for this seemingly bizarre proposal, so much so that the then editor of *Nature*, the world's premier scientific journal, was moved to suggest that Sheldrake's was a book fit for burning.

I was sufficiently troubled by this suggestion that I incautiously suggested to Sheldrake that he and I do a joint experiment, based on the behaviour of my chicks, to test his idea. We agreed the design of the experiment and made two rival predictions as to its outcome, and decided that when it was done we would write up the results as a joint research paper. Within my paradigm, the predicted outcome of the experiment would be that the behaviour of successive hatches of chicks would not change, despite previous hatches having had a novel experience; within his, there should be a change, as later hatches would acquire a memory of the experience of the earlier hatches by virtue of some incorporeal 'morphic resonance'. When we ran the experiment, I was proved right – to my satisfaction and to that of other researchers in the field. Sheldrake, however, was able to convince himself that, viewed in a particular way, the data supported his hypothesis of morphic resonance. We couldn't agree on how to write the joint paper, and instead published two alternative accounts side by side.[8] This just goes to show how little facts 'speak for themselves'. We all cling tenaciously to our views of the world; rather than accept an interpretation which destroys our paradigm, we wrap the paradigm in supplementary hypotheses. The history of attempts to prove or disprove extrasensory perception and related phenomena shows many similar episodes.

WHERE DO PARADIGMS COME FROM?

The most interesting consequence of Kuhn's work, however – and this probably despite his original intentions – was that attempts to understand the nature of scientific knowledge were wrested from the hands of abstract philosophers, and opened up to the growing number of sociologists concerned with what has become known as the sociology of scientific knowledge. They could, and did, ask the question Kuhn apparently never thought of asking: where do our paradigms come from? Kuhn himself, who spent most of his time from the publication of *The Structure of Scientific Revolutions* in 1962 till his death in 1996 studying the history of physics, seems to have taken it for granted that paradigms emerged as a result of an accumulation of theoretical problems within a particular science. But if paradigms are not absolutely determined by 'the facts' of science, then our reasons for preferring one to another must include factors outside science – such as religion, social expectations or ideology. Thus the claim that science produces 'truth' about the world is forced even further on the defensive. Facts are not merely at the disposal of provisional hypotheses to account for them, but the whole way in which we view and interpret them can now be re-patterned simply by shaking the paradigm-kaleidoscope. The consequences were startling. Within the philosophy of science itself, even Popper's erstwhile pupils abandoned his view of hypothesis-making. For some, what mattered became simply whether any particular 'research programme' was productive or had become sterile or degenerate.[9] For others there was no longer any such thing as scientific method; what worked, worked.[10]

Kuhn had thus dug a tunnel below the seemingly impregnable fortifications of natural science. This allowed the return of a quite different view of what drove science forward, deriving from the writings of Marx and Engels a century previously, but made explicit in a famous international meeting on the history of science held in London in 1931.[11] At that meeting, an unanticipated delegation arrived from the still relatively young Soviet Union, headed by the powerful Marxist politician and theoretician Nikolai Bukharin (later purged and shot by Stalin). The key paper was entitled 'The social and economic roots of Newton's *Principia*', and was delivered by Boris Hessen. In it he argued that, far from being a work of pure scientific scholarship isolated from the social conditions of the time, Newton's experiments, theories and the framework in which they were set – their paradigms therefore, in Kuhnian language – had been shaped by the new economic demands of England's rising merchant classes. The merchants needed accurate navigational tools for the ships which carried the imports and exports on which the Industrial Revolution would be built; first Galileo, and then Newton, provided these through the new mechanics and cosmology their work created. Hessen went on to trace the subsequent history of the physical sciences through the nineteenth century, linking them to both the economic needs and the ideological commitments of emergent capitalism.

Here, then, was a quite different way to think about the growth of science. It gave impetus to a burst of Marxist scholarship in the 1930s, subsequently to be submerged in the aftermath of Stalin's brutal and bloody imposition of a dogmatic and sterile orthodoxy on Soviet science, and the Cold War of the late 1940s and 1950s.[12] Knowingly or not, Kuhn reopened this line of analysis in the West. Sociologists of science, and the critical 'social responsibility of science' movement of the late 1960s and early 1970s, began to look at the relationship between dominant scientific paradigms – or at least theories and metaphors – and ideas about economics and society.

The sources of our paradigms in the biological sciences seem particularly sensitive to such social, economic and cultural inputs. As I pointed out in the last chapter, much scientific argument and hypothesis-making proceeds through the use of analogy and metaphor. This is especially true in biology, perhaps because the subject matter of biology is so difficult, perhaps because of our deference to physics and technology. Whatever the reason, we often use metaphors derived from simpler sciences, which we believe we understand better, to conceptualize our subject – that is, to create its paradigms. I have already introduced three such metaphors: the heart as a pump, brain memory as computer memory, and ATP as the cell's banking system. The first two are derived from human artefacts, the third from a key feature of the organization of industrial societies. As I have already hinted, the temptation to rely on mechanical and industrial metaphors for living processes goes back to the transformation in scientific thinking that came with the Newtonian revolution of the seventeenth century, itself of course intimately connected with the birth of modern capitalism and industrialization. Before that time, the metaphor trade tended to be in the opposite direction: the physical worlds of our own Earth and the cosmological universe were described in language usually reserved for living organisms, as when inanimate forces (the wind, rivers, and so forth) were ascribed intentions and goals.[13] The significance of this reversal cannot be overestimated, for with it came the birth of the reductionist methodology which has so influenced biological thinking in the subsequent three centuries.

Metaphors help us think about our subject – but they may also be a hindrance, for they also constrain the way we think.[14] In the biochemical literature from the 1930s through to the end of the 1950s, cells were pictured as small factories, with 'powerhouses' (mitochondria) and energy currency systems (ATP), whose central function was maintaining a balanced energy budget. From the 1950s on, a subtle change in metaphor is discernible, and by the 1980s energy budgets had been relegated to a minor league. Dominant now were concepts of control processes and information flow within the cell, whose functions were seen no longer in terms of crude energy, but of sophisticated management. DNA and RNA, and to a lesser extent proteins, became grouped together as 'informational macromolecules', which is how you will find them in many standard biochemistry and molecular biology textbooks of the present day. I shall explore the implications of this metaphor more fully in Chapter 5.

The coincidence in time of this switch with the change in how society as a whole has come to view the central issues of its economy is too striking to pass over. We are told that we are an 'information-rich society', and that our bodily processes too are centrally concerned with the management, reproduction and transmission of information. Brains, once perceived as functioning on hydraulic principles, and later as telephone exchanges, are now supercomputers, another part of biology's information superhighway.

Such metaphors are more than merely easy ways to make complex phenomena comprehensible. Sociobiological analysis in the hands of E. O. Wilson and others employs identical mathematical models to those used by a particular school of monetarist economists based in Chicago (and the compliment is returned by economists who have created a new discipline called 'evolutionary economics').[15] Monetarism is more than merely a theory of economics which was contentious in the 1970s, became a cornerstone of Thatcherism and Reaganomics in the 1980s and now, surrounded by the wreckage of the economies it has destroyed, is largely discredited. More than that, it depends centrally on a reductionist view of society which coincides precisely with the sociobiological approach to both human and non-human animal behaviour.[16] In the 1990s the metaphors are changing again: chaos theory is now applied to predicting Stock Exchange fluctuations as well as to population dynamics in complex ecosystems.[17]

Because of the deference paid to biology over 'softer' social sciences, such paradigm-coincidences had and continue to have a social utility, in ways which I shall explore in more detail in later chapters. For instance, the re-expression of old theories about differences in IQ between blacks and whites in the USA, or of 'the inevitability of patriarchy', have coincided with the backlash in the 1970s and early 1980s against the black and women's liberation movements.[18] Science, we have learned, is not 'neutral'. Its objectivity is only skin-deep, as the paradigm on which our theory-building and observations are based is shaped at least in part by our own social expectations and philosophy. Feminist biologists, philosophers and historians of science in particular have been quick to point out how far the science we do and the paradigms within which we work – the very ways in which we see and interpret the world around us – are shaped by the gender expectations of science as still an overwhelmingly masculine activity. The clearest-cut examples come from the field of animal behaviour, where feminist sociologists and historians of science have been able to document how men researchers on animal behaviour observe and record what they regard as the significant behaviours of the animal groups they study quite differently from the way that women (let alone feminist) researchers do.[19]

PERFORMANCE VERSUS TRUTH

This opening up to question of the very status of scientific knowledge has led in the 1980s and 1990s to fierce battles within both the philosophy and sociology of science, between opposing camps. On the one side there are 'relativists', who argue (I simplify) that there are many ways of describing the world and that modern science, itself a cultural construct, has no superior claim to 'truth'. On the other side, the 'realists' maintain that the scientific method can provide some approximation to true knowledge of the material world. These battles have led to charges of 'political incorrectness' and of 'assaults on reason' which go far beyond my concerns here.[20] I wish to consider just one aspect of the debate: the relativists' charge that science tells but one story among many possible about the world. Defenders of traditional views of science respond that, after all, science and technology work: for example, aeroplanes, designed according to the most rigorous principles of physics and engineering, don't fall out of the sky. But that a piece of science or technology works does not imply that the theory on which it is based is necessarily true. Melanesians apparently navigate their canoes and make accurate landfalls on islands many travelling days distant by regarding the sea as moving past them while their boats and the stars by which they navigate remain stationary. Some physicists at least have no problem with this. Thus Stephen Hawking, in debate with mathematician Roger Penrose, whom he attacks as a Platonist, states bluntly enough:[21]

I take the positivist viewpoint that a physical theory is just a mathematical model and that it is meaningless to ask whether it corresponds to reality. All that one can ask is that its predictions should be in agreement with observation.

A few years ago I was approached by Art Janov, the founder of a form of psychotherapy known as Primal Screaming. Janov was convinced of the validity of the theory on which his therapeutic method, a form of 'rebirthing', was based, and was further convinced that depressed clients who underwent his therapy should show biochemical and immunological changes which indicated that they were improving. Could I test this idea? I agreed to make measurements on blood samples taken from the clients both before and up to a year after they went through their screaming therapy. One of the measures I chose was of the quantity of receptor molecules for the neurotransmitter[22] serotonin present on the surface membranes of a particular class of blood cells (platelets). These are the receptors which are the target for the class of drugs known as selective serotonin re-uptake inhibitors (SSRIs); Prozac is one of the best-known examples.

As Janov had hoped, it turned out that before therapy, the quantity of these particular receptors in his clients' platelets was considerably below normal. Within six months of therapy, clients' depression had lifted, and the biochemical and immunological measures I was making approached the average for 'normal' non-depressed people of the same age and sex. Janov was (and I believe still is, for he has cited this finding in books he has subsequently written[23]) convinced that this proved his therapeutic theory to be valid. But while there is a weak correlation between the biochemical measures I was making and standard psychiatric rating scales for depression, there is no way of knowing whether (a) his clients would have recovered even without therapy, or more importantly, (b) whether the therapy Janov offers works because his theory about it is correct, or because he is a charismatic figure whose clients recover because they believe that they will get better if they scream appropriately. Indeed, I obtained similar biochemical results when, a couple of years later, I did a similar study with depressed clients going through other, less dramatic forms of psychotherapy, so I suspect that in such cases the therapist matters more than the therapeutic theory.[24] The therapies thus meet the criterion that they 'work', inasmuch as clients going through them show behavioural and biochemical changes in accordance with prediction.

However, these changes apparently occur irrespective of the therapeutic theories on which the treatments are based.

A second example. According to official US figures, up to 10 per cent of all American children – mainly boys aged between 8 and 14 – are currently being diagnosed as suffering from a condition known as ADHD – attention deficit hyperactivity disorder. The criteria for making this diagnosis centre on the child's school performance. ADHD children are said to be inattentive and a nuisance in class, unable to sit still or accept the authority of their teacher – and sometimes of their parents. Once such a diagnosis has been made, the recommended treatment is to give the child a psychotropic drug, an amphetamine-type substance called ritalin, which is believed to act upon neurotransmitters in the brain.[25] Although in a small number of the cases in which the drug is used there may be some unusual level of activity of the neurotransmitter or its receptors in the brain, for most children this is both unknown and unlikely – and in any event, the behavioural significance for a person who has levels of neurotransmitter which differ from the average for the population is simply unknown.

Neither the diagnosis, nor the treatment with ritalin, is recognized to any significant degree outside the USA, although as I write (1996) determined efforts are being made by a few psychiatrists and parents to bring it to the UK, with considerable attendant publicity. There is no doubt that ritalin sedates such children and makes them more tractable at school – indeed, ADHD seems to be a disorder that remits at weekends and school holidays, at which times the drug often appears unnecessary. So ritalin also 'works' – that is, it makes the children taking it easier for their teachers and parents to handle. But the theory on which it is based – that the problem for which it is being prescribed lies 'inside' the child's brain, rather than in aspects of his relations with his parents, the ability of his teachers, the size of the class in which he is being taught, or the social relationships within which he is growing up – is almost certainly wholly fallacious for the vast majority of cases in which the drug is being used.

TECHNOLOGY

There is yet another factor to take into account in defining the powers and limits of science – technology. The conventional distinction between the two is that science provides knowledge of the world while technology offers the power to manipulate it. I am unhappy with this distinction for several reasons which need not concern us unduly, as they are not relevant to my purpose here. All I need say is that the distinction is artificial – perhaps connected with the traditional British valuation of head-work over hand-work. From my perspective, one person's science is another's technology. The Apple Mac on which I key in these words is to me a piece of technology; I use it without caring how its mouse and hard disk work. Yet the efficient and trouble-free functioning of my computer depends on the science of the mathematicians, computer scientists and engineers who designed and built it or developed programs for it. Equally, the chick memory research I publish is science for me, but technology for someone who wishes to use the chicks' behaviour as a means of testing a new memory-altering drug. In many areas of modern molecular biology the distinction is even less clear, and we now have the term 'technoscience'.

The simple observations which I described at the start of Chapter 2 required little more than my own unaided senses, a watch, and notepad and pencil. Admittedly, I introduced the possibility of a bit of automation – a video camera, a sensor in the bead, some computing power to work out the time budgets of an ethogram. But if I had taken you only one step further down my experimental path, you would have met some very big machines indeed: centrifuges capable of generating a force half a million times as strong as gravity; electron microscopes powerful enough to enlarge your thumbnail to 5 kilometres across; gene synthesizers which can link together defined sequences of nucleotide bases into simulacra of natural DNA with the insouciance of an experienced knitter . . . and backing them all up, a powerful industry of instrument manufacturers and chemical companies producing, for a price, anything and everything my laboratory might require.

Without such instrumentation and industrial support systems, no modern biology lab could survive. It is not that the questions we ask about the living processes we study are merely not answerable without the technology, they are literally unthinkable. Before the development of effective lenses, and then optical microscopes, in the seventeenth century, the existence of the overwhelming majority of the living world – the bacteria and other single-celled organisms which constitute so much of the planet's biomass – was wholly unsuspected. Antony van Leeuwenhoek's drawings (Figure 3.1) of the 'animalcules' his

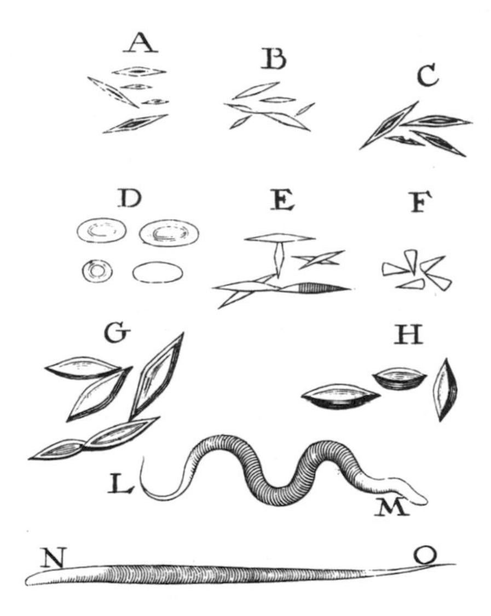

Figure 3.1 Antony van Leeuwenhoek's single-celled 'animalcules'.

microscope revealed in a drop of pondwater, publicized in the 1670s, revolutionized biology to an even greater degree than Galileo achieved for cosmology when he turned his matching device, the telescope, to the heavens and observed the moons of Jupiter. Until then, known living forms were pretty much limited to those with which the author of the Book of Genesis had populated Noah's Ark.

The fact that even those living forms that were known prior to the microscope are composed of *cells*, which biologists now regard as the basic units of life, went undetected and unsuspected. It was not until the middle of the nineteenth century that, on the basis of microscope studies, Matthias Schlieden, in Germany, concluded that all plant tissues were composed of cells. (The name 'cell', derived as a metaphor for monks' cells, was in fact coined by Robert Hooke, a century and a half previously, to describe the dead structures he found in cork.) Theodor Schwann came to similar conclusions about animal tissues. As a result he was able to make the comprehensive claim that all organisms consist of one or more cells, and that the cell is the basic

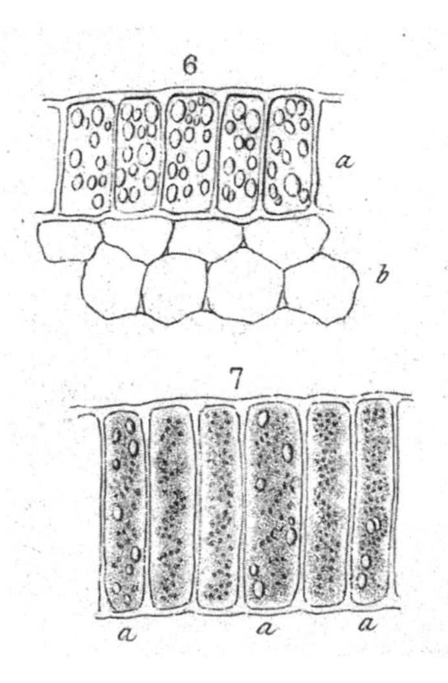

Figure 3.2 Theodor Schwann's drawings of cells.

unit of structure for living organisms (Figure 3.2). Then the English botanist Robert Brown identified a small round structure within all the plant cells he examined, and called it the nucleus. Similar structures were soon found in all multicellular animals and plants. They were present too in certain single-celled organisms among van Leeuwenhoek's zoo, but missing from others. Cells with nuclei became known as *eukaryotes*; those without, which include bacteria, are *prokaryotes*.

The microscope thus opened up a new world, not merely of free-living 'animalicules' but also of the cellular and subcellular structures within the tissues of plants and animals. None the less, there are strict limits to the magnifying power of even the best of optical microscopes. Until the advent of the electron microscope in the early 1950s, the internal constituents of cells such as mitochondria were unobservable and hence unknown. It was impossible either to build theories about the partition of cellular functions which such subcellular particles might embody, or to develop the science and technology of the centrifuge which might enable one to separate these structures from the fluid – the cell's *cytoplasm* – in which they float.

Technology – by which in this sense I mean available instrumentation and methods – both solves certain problems and suggests others. But just as the cage constrains the marmosets or chicks, so does technology also constrain the way we view the world. Take electron microscopy as an example. In order to prepare living tissues to be viewed through such a microscope, it is necessary first to 'fix' them in a solution which pickles the cellular constituents. The next step is to embed the small fixed piece of tissue in a resin such as Araldite, and stain it with a chemical which binds selectively to certain cellular components (lipids, perhaps, or proteins, or even very specific types of protein) and which can be made opaque to the electrons with which the sample will in due course be bombarded. Finally, it is necessary to cut very thin slices of the sample, perhaps no more than a thousandth of a millimetre thick, place these slices on a copper grid, and insert the grid into a vacuum tube, ready to be pounded with electrons. A picture of a cell produced by such a process is shown in Figure 3.3.

Figure 3.3 illustrates very well the extent to which one has to learn to see the patterns of differing shades of grey as 'representing' cells, their nuclei, mitochondria, membranes and so on. To the novice these patterns make little sense. The apprentice electron microscopist is taught just how and what to see, what to regard as 'real' and what as 'artefact' – the unwanted consequences of one or more of the procedures used to prepare the living tissue. Thus the new observer is initiated into the conventional wisdom developed by half a century of biological work in the artificial world of electron microscopy.

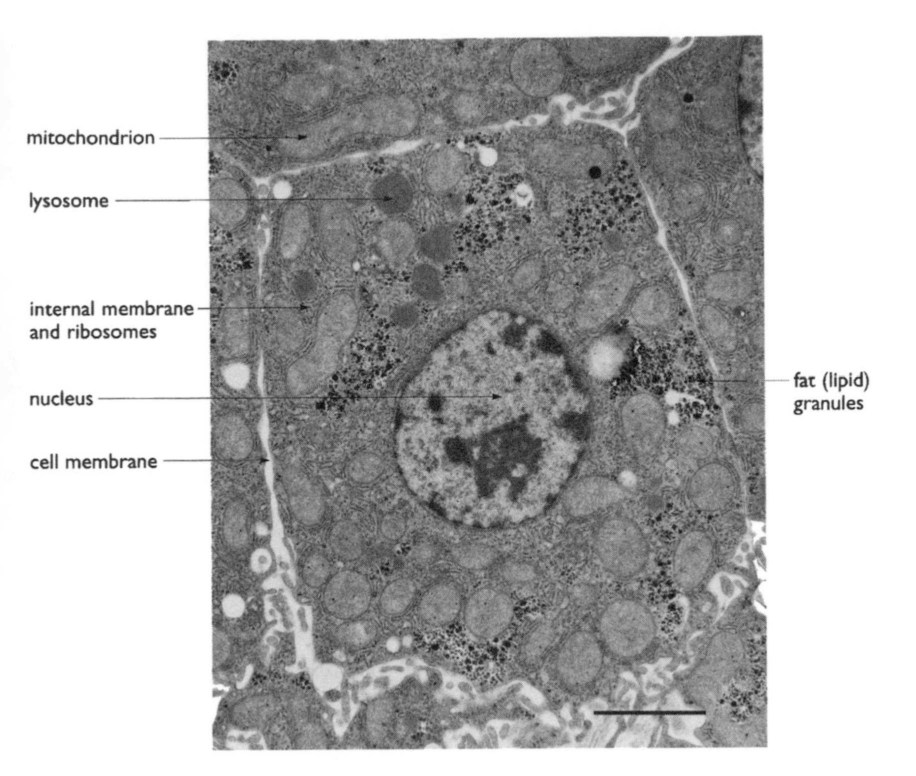

Figure 3.3 Electron micrograph of a liver cell from a young chick. Scale bar 1×10^{-6} (or 1 micron, μm).

Other techniques, such as video recording under a different form of microscopy, called phase-contrast, show living animal cells to be three-dimensional, dynamic structures, in constant interaction with their environment and filled with complex internal particles such as mitochondria which are themselves not static but in continuous movement. Their components are not the black and grey, one- and two-dimensional, static patterns seen in Figure 3.3. Yet it is these fixed patterns of the electron micrograph that, as a result of the technology, form the basis for drawings of cells in biology textbooks, and provide the conventional 'mind-picture' of cells even for experienced

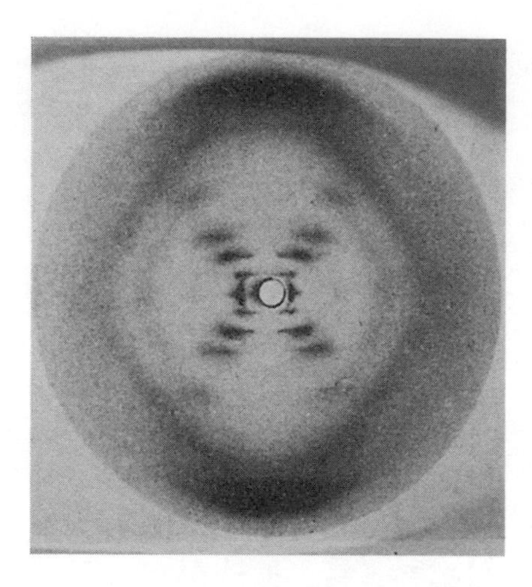

Figure 3.4 Rosalind Franklin's X-ray diffraction pattern of a DNA crystal.

biologists. So powerful is the technology that it becomes very hard to move beyond it, to think in three, let alone four dimensions.

This problem is certainly not confined to 'visualizing' cells microscopically.[26] Consider, for instance, Figures 3.4 and 3.5, and, if you know no biology, try to guess their meaning. Each is a pattern of black and grey smudges on a whitish or paler grey background. (This is not a consequence of the need to reproduce coloured images in black and white; they are indeed the patterns biologists have to learn to interpret.) Figure 3.4 is a diffraction pattern, obtained by directing a beam of X-rays through a carefully aligned crystal and allowing the scattered beam to hit a photographic plate. And not just any old crystal, either: you are looking at the pattern Rosalind Franklin obtained when she was studying DNA, and from which she was able to deduce its double-helix structure – the observation so brilliantly exploited by James Watson and Francis Crick. How does the pattern reveal the structure? Well, it is all to do with the number of spots and their spacing – if you observe them with the eye of love and the benefit of experience.

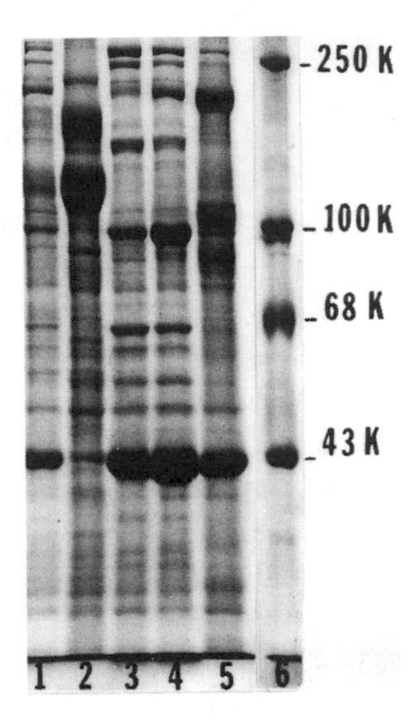

Figure 3.5 Proteins separated on a gel. Each of the first 5 'ladders' separates a different starting group of proteins. The sixth contains molecular weight markers. K = 1000 and refers to molecular weight.

Figure 3.5 shows the result of an analysis of proteins present in particular brain cells. In the analysis, protein solutions were placed on a slab of a gel-like substance, and an electric current was passed through the slab. This caused the proteins to be driven down through the gel at a rate dependent on their individual electrical properties and molecular weights. After a few hours the current was switched off and the gel bathed in a solution which stained the proteins. Each 'rung' of the ladder-like patterns visible in Figure 3.5 represents a different group of proteins, which can then be cut out of the gel and studied in isolation. From the gel, the initiated can interpret the molecular weights and relative concentrations of different proteins, how fast they are being synthesized in the cell, and many other features as well. The technology is relatively simple, but without it the world it reveals would not exist to the researcher. Indeed, it does not even exist in this form in the cell at all; to produce it the proteins have been subjected to procedures which destroy and degrade them (they have first to be precipitated, and boiled in detergent solutions); whatever their status as 'natural kinds' within the cells from which they have been derived, by the time we come to observe and study them by this technology they are no longer what they were.

These are examples of the sciences made possible by technology, the technologies made possible by science. The world-view we biological researchers create is derived from the intimate interaction of technology and science with the eye of craft experience, shaped by the theoretical expectations according to which we operate. It is a world which presents challenges which go deeper than Popperian hypothesis-making, Kuhnian paradigms, and truth-versus-performance arguments by which those studying the epistemology of science attempt to make their own sense of what we do. Wresting reliable knowledge from the world we biologists study is, as the novelist Arthur Koestler once described it, an Act of Creation.[27]

The shapes, patterns and structures we see through an electron microscope are artificial. They have been created by the complex of procedures through which the living material has been transmuted. Much of what modern science does, the problems with which it concerns itself, is thus divorced by technology from the immediately observable world, indeed is literally created – the product of human labour.[28] In the 1970s, anthropologist and sociologist Bruno Latour spent a period living with the strange tribe of biochemists working in the prestigious Salk Institute in California. The tribe he studied was locked into a 'race' with a rival laboratory to discover the structure and biological activity of a peptide hormone. Latour recorded the way in which the researchers talked about how they created knowledge from the machinery of discovery within which they were immersed; how yesterday the molecule had one structure, today – as a result of a new measurement – it had another. Since I read Latour, I have become much more aware of how we biologists speak of the objects of their (our) study; not as if yesterday we *thought* one thing, and today we *know* better, but rather as if the change were in the 'real' world outside us; yesterday this world took a particular form, while today it takes another. Not 'we now know that there are eight different forms of the metabolic glutamate receptor, whereas last year we knew of only seven', but 'there *are* now eight different metabolic glutamate receptors'. We speak as if the world inside our heads had primacy over the natural world outside, even if we indignantly reject such a charge when it is laid against us.[29]

SO HOW DO WE KNOW WHAT WE KNOW?

Given all that I have said, what justification can there be for claiming that it is possible to obtain reliable knowledge of the living world, and that biologists can achieve such knowledge? There is no non-circular way to answer that question, so let me acknowledge the circularity and begin with some biological assertions made from within an evolutionary perspective. It is by making these assertions that I wish to retain my claim to be able to draw conclusions about the nature of the living world which, while not immune to the critique of sociologists or philosophers, none the less approximates to how the world *really* is.

The evolutionary lineage which led to humans has been characterized by the development of more flexible organisms with bigger and more powerful brains, able to adapt to widely differing conditions and respond to rapidly changing circumstances. As J. B. S. Haldane pointed out, no other animal can run ten kilometres, swim two, and then climb a tree; I would add, let alone sit down and write about it! As will become clear in later chapters, this is not the only way to succeed evolutionarily, nor necessarily the best, but it is the way which, on the available evidence, led to humans. Human survival as a species is not based on our ability to outrun a predatory carnivore or even a potential prey, to roll into a ball protected by spines or a hard carapace, to inform potential enemies that we are dangerous or distasteful by bright coloration, to camouflage our appearance or to hide in burrows, coming out cautiously only after dark. To survive and succeed despite our incapacity to do these things, we – and our immediate evolutionary forebears – have had to rely above all on our brains.

And what our brains permit us to do above all is to have foresight – to think ahead, to predict the consequences of our actions and those of others around us. If those predictions were mistaken, we would not live for long. That is, human survival depends on our being able, as a species, to make relatively reliable hypotheses about the world around us and to act appropriately upon those hypotheses. Among other things, therefore, humans are hypothesis-makers (and were so, long before Popper pointed it out). Certainly, these hypotheses, and the observations upon which they are based, are actively constructed rather than passively received. But if the mental world we construct in this way did not correspond reasonably accurately to the way the world outside 'really' is, we could not survive. A hypothesis that the vehicle we see rapidly approaching us as we cross the road is an optical illusion or is made of pink marshmallow is unlikely to ensure the longevity of the hypothesis-maker.

Such hypothesis-making may be seen as the starting-point for science. It has been claimed that in this sense all animals 'do' science, in that most species with brains and nervous systems a little more complex (in terms of numbers of cells and of the connections between them) than flatworms can learn and generalize from experience.[30] But science of course is much more than this: it is *socially organized* hypothesis-making. That is, to be scientific, hypotheses must be shared, tested and eventually agreed among a community, a sharing and testing that diminishes the chance that they are the idiosyncratic consequences of a particularly unusual brain at work.

This sharing is both the strength of science and its weakness. Without it, nothing would be possible. Hypotheses that a chick or a marmoset makes and the generalizations that it draws from them die with the animal's own death. Even for most relatively large-brained animals, each individual, in each generation, must hypothesize for itself, and each individual in the next generation must start again as if nothing had ever been learned. Even if it watches closely, one chick cannot benefit from another's experience of the bitter bead and therefore refuse to peck: it has to taste the bead for itself.[31] However, social animals with large brains, such as monkeys, can learn from the experience of others, and there have even been claims of cultural transmission across generations. An example often cited derives from a study of a particular troop in a semi-wild colony of macaque monkeys in Japan in the 1950s. The monkeys were given dirty sweet potatoes as food; one began to wash the dirt off the potatoes in a stream (it subsequently generalized, and washed corn too). Over the next six years the potato-washing habit was said to have spread to more than half the troop, with parents teaching their offspring.[32] The simple interpretation found in most textbooks, that this represents a form of social learning and cultural transmission, has however been thrown into doubt by more recent research which has shown that such 'washing' behaviour is probably associated with thirst and acquired spontaneously,[33] without the need to be 'taught'.[34] But even if the findings were uncontroversial, nothing that occurs in the non-human animal world matches the cumulative nature of hypothesis-making that constitutes human science. We are able to build on the tested and seemingly validated hypotheses not merely of those currently alive, but of all previous generations. This capacity must have been immeasurably strengthened once an oral cultural tradition was superseded by a written one, enabling records of past hypotheses and tests to be preserved.

However, humans are more than just scientific hypothesis-makers. We live in communities shaped by many other cultural and economic forces, forces which provide strong guidance as to how we should view the world around us and our fellow-humans. In Britain in the 1990s, where the gap between rich and poor is greater than it has been for a generation and is still increasing, the directors of the newly privatized utilities and the people they have sacked and rendered unemployed see the world from very different perspectives. In a society in which there is a strong division of labour and power between men and women in every field of work from science to child care, their viewpoints on the world will also differ. A white racist football fan is unlikely to make the same hypotheses about the world as the black player he abuses.

For many fields of scientific hypothesis-making, these rather crudely drawn distinctions may be irrelevant. They may not affect cosmology – although a person's religious perspective is certainly likely to. They may not affect physics or chemistry – although they may well affect one's attitude to nuclear power or ozone depletion. But biology is different. Not only is the living world much more complex and less predictable than the inanimate world studied by physicists and chemists, but biology, as I continue to emphasize, lays claims to be in a position to tell us, as humans, who we are, where we came from, where we are going, how we must live and relate to our fellow living creatures. It does what religion used to do. Indeed, as will become clearer in later chapters, biology has explicitly taken on this role ever since Darwin. This is powerful stuff, and so it should not surprise us to find that ideological preconceptions are becoming more apparent, and shaping our hypotheses more decisively. Above all, hypotheses are about the joints at which to carve nature. (Is this phenomenon or process I am observing an example of aggression, or of memory, or of play? Or at another level, of a protein, an enzyme, or whatever?) They depend on that seductive but misleading trio of metaphor, analogy and homology. If we use metaphor as if it were analogy or homology – the brain *is* a computer, DNA *is* a code – or if we use analogy as if it were homology – for instance by claiming that two animals biting and clawing at each other are behaving in a way homologous to human 'aggression' – then we delude ourselves.

The metaphors and analogies we find attractive are laden with cultural values and expectations that come from outside our science. They inevitably reflect our experience as directors of companies or as sacked workers, as men executives or women child-carers, as white racists or black footballers. That is, they are not and cannot be free from ideology. Those who deny this – and there are many among biology's leading ideologues who claim to have purged themselves of such vices, able to go about their work by holding a perfectly reflecting mirror up to nature – are at best unselfreflective. Even more inexcusably, they are wilfully and woefully ignorant of the hard work done by philosophers and sociologists in developing an understanding of the nature of science and the knowledge it creates.

Despite this doubt at the very core of the scientific endeavour, we are not in a position to assert that 'anything goes'. Although the observations we make about the world are theory- and ideology-laden before we start, and the joints into which we carve nature are provided less by *a priori* definitions than by operational need, they must still make a reasonably good fit with the world, or we could not proceed. Our hypotheses would fail. Science-fiction writers, but not scientists, are permitted to create life by harnessing the power of lightning and passing it through assembled fragments of corpse. The monster is Mary Shelley's creation, not Dr Frankenstein's. And, as again will become clear later, however great their budget, genetic engineers will not be able to turn humans into angels, nor cryogenicists restore the memories of the past owner of a severed and deep-frozen head.

TAKING STOCK

Where has this Cook's Tour through several decades of intense debate about the nature of scientific knowledge got us? I began the previous chapter by describing what seemed at first to be straightforward observations about animal behaviour. The moment these observations were analysed in greater depth, however, even the simplest statement about the natural world seemed to be built on foundations of shifting sand. Philosophical and sociological critics have succeeded in diminishing many of science's claims that it provides the methods by which truth about the material world can be obtained. The phenomena we describe and purport to explain appear to be constructed according to hypotheses derived from our own fallible senses, cultural traditions, social expectations and limited technological powers. And yet, science and technology seem to do more than just work. They offer us more than mere power to manipulate the universe; their claims to provide reliable knowledge are surely better grounded than those of cults and religions. I maintain that we can and indeed must accept the strength of the philosophical and sociological critiques and that we can yet save our science's claims to (constrained) reliability. Culture-bound and shaped by technology our hypotheses may be, yet they are constantly confronted by a reality test. We can no longer maintain that the Earth is flat, the Moon is made of green cheese, or that IQ tests measure some fixed, biologically determined feature of an individual. Science has shown us better. Indeed, if I answered otherwise I could scarcely continue to work as an experimental scientist – a naïve realist, as it might be described. None the less, I insist that my claims for science are more than merely special pleading by someone who would lose both their job and their life's work were it otherwise.

At any one time, the science we do, the questions we ask about the world, the hypotheses we frame and the answers we find satisfying, depend on a constant interplay of factors. They include what is sometimes called the internal logic of the subject – that is, the cumulative state of knowledge or belief about the particular problem or question as currently understood by the community of researchers interested in it – in short, their paradigms. But they also include the current state of technology. It was no good asking a question which demanded knowing the amino acid sequence of a protein in the 1940s, before any protein had ever been sequenced, still less before there were machines which could perform the operation routinely and rapidly. It is no good asking a question which permits an error of less than 1 per cent if the equipment one is using is theoretically and practically incapable of better than plus or minus 10 per cent. Getting these two both right is what Medawar called the art of the soluble.

But what neither Kuhn nor Medawar allow, and on which philosophers, sociologists and social critics of science insist, is the external framing of our subject. This includes the economic and political logic which drives society to fund some types of research and not others, and more subtly it includes the cultural and social forces which shape our metaphors, constrain our analogies and provide the foundations for our theories and hypothesis-making. It is these forces that have helped drive forward biology's currently dominant reductionist mode of thinking, and which a more comprehensive science must now transcend.

NOTES

1. Karl Popper, *The Logic of Scientific Discovery.*
2. Government White Paper, *Realizing Our Potential.*
3. Stephen Jay Gould, *Wonderful Life.*

4. Thomas S. Kuhn, *The Structure of Scientific Revolutions.*
5. Daniel C. Dennett, *Darwin's Dangerous Idea.*
6 Robert Olby, *The Path to the Double Helix.*
7. Rupert Sheldrake, *A New Science of Life.*
8. Sheldrake, 'An experimental test of the hypothesis of formative causation'; Rose, 'So-called "formative causation" '; Sheldrake, 'Rose refuted'.
9. Imre Lakatos and Alan Musgrave, *Criticism and the Growth of Knowledge.*
10. 'Anything goes,' argued Paul Feyerabend in *Farewell to Reason*, although anything apparently did not include criticizing Feyerabend himself, which evoked an apoplectic response: see Feyerabend, *Killing Time.*
11. Nikolai Bukharin *et al.*, *Science at the Crossroads.*
12. Hilary Rose and Steven Rose (eds), *The Radicalisation of Science.*
13. M. A. Hoskin, in his Introduction to the 1961 edition of Stephen Hales' *Vegetable Staticks.*
14. Hermann Haken, Anders Karlqvist and Uno Svedin (eds), *The Machine as Metaphor and Tool.*
15. J. Hirschleifer, 'Economics from a biological viewpoint'.
16. Richard C. Lewontin, *Biology as Ideology.*
17. Stuart Kauffman, *At Home in the Universe*; Peter Coveney and Roger Highfield, *Frontiers of Complexity.* It is interesting that the British government's Chief Scientific Adviser, Robert May, appointed in 1995, is a world expert on chaos theory as applied to anything from ecosystems to markets.
18. Rose, Lewontin and Kamin, *Not in Our Genes.*
19. Donna Haraway, *Primate Visions.*
20. Andrew Ross (ed.), *Science Wars*, especially the chapter by Hilary Rose, 'My enemy's enemy is – only perhaps – my friend', pp. 80–101.
21. Stephen Hawking and Roger Penrose, *The Nature of Space and Time*, pp. 3–4.
22. *Neurotransmitters* are the chemical messengers that carry signals between one cell and the next, or between nerves and muscles. For rather obscure embryological reasons, receptors for these messengers are also present in a particular class of cells found in the bloodstream, called *platelets.*
23. Arthur Janov, *The New Primal Scream.*
24. Sarah Willis, 'The influence of psychotherapy and depression . . .'.
25. Drug and Chemical Evaluation Section, 'Methylphenidate: A background paper'.
26. This standard technical term is itself quite revealing, as it indicates much more than just seeing or observing, but instead implies literally making visible that which is otherwise invisible. The biochemist Bill Pirie, commenting on

the draft of this chapter, suggested using the term *iconize* instead, to point yet more strongly to the hidden implications of the language.

27. Arthur Koestler, *The Act of Creation.*

28. Jerome R. Ravetz, *Scientific Knowledge and Its Social Problems.*

29. Bruno Latour and Steve Woolgar, *Laboratory Life.*

30. See Robin Dunbar, *The Trouble with Science*; the arguments in this book are somewhat muddled, however.

31. We have recently discovered that this is not quite true. Amy Johnston and Tom Burne, post docs in my lab, have found that 'observer' chicks, which see their colleagues pecking but are prevented from doing so themselves, will also tend to avoid dry beads later. A. N. Johnston, T. Burne and S. P. R. Rose, *Animal Behaviour* (in press).

32. M. Kawai, 'Newly acquired precultural behaviour of the natural troop of Japanese monkeys on Koshima islet'.

33. Elisabetta Visalberghi and Dorothy M. Fragaszy, 'Pedagogy and imitation in monkeys . . .'.

34. R. W. Byrne, 'The evolution of intelligence'.

4

The Triumph of Reductionism?

. . . we are assured that the all-wise Creator has observed the most exact proportions, of number, weight and measure, in the make of all things; the most likely way therefore, to get any insight into the nature of those parts of the creation, which come within our observation, must in all reason be to number, weigh and measure. Stephen Hales, *Vegetable Staticks*

THE CRITIQUE OF REDUCTIONISM

The time has come to get to grips with the issue of reductionism in biology. The term itself is a source of much polemic. To some, it is an unqualified boo-word, representing a way of emptying life of its manifold rich meanings, of turning individual personal experience into chemistry and physics, mere mechanisms. This search for other meanings lies at the heart of New Age philosophy's rejection of reductionism – a rejection abetted by a few ex-biologists, well exemplified by Rupert Sheldrake and his theories of 'morphic resonance' – indeed, I can think of no one who better fits Dawkins' epithet 'holistier than thou'. For others, however, the critique is systematic and based upon a coherent philosophical and political analysis which sees modern science as the inheritor of nineteenth-century mechanical materialism, itself tightly linked ideologically to a particular phase of the development of industrial capitalism. This is the case that I, Lewontin and Kamin argued in *Not in Our Genes*. Other forms of the critique are advanced by feminist philosophers of science for whom reductionism typifies the limited rationality of the peculiarly masculine, cognitive, objectifying approach to the world taken by modern science, with its concomitant refusal to respect the validity of subjective experience.[1] In similar vein, some ecologists criticize reductionism because it appears to deny the interconnectedness of phenomena. Because it fails to understand the unity of life, of the Earth as living Gaia, it is as a result dangerously liable, either by advertence or inadvertence, to destroy the planet.[2] There is, of course, substance to such critiques, but they are not of prime concern to me here. In this chapter I want first to consider both why reductionism has been and continues to be a powerful scientific method, and therefore attractive to many biologists, and also why it is ultimately unable to answer many of the most fundamental questions with which biology is concerned.

POPPER VERSUS PERUTZ

Despite the vigorous tradition of non-reductive thought in biology, a tradition whose lineaments will become clearer in the pages that follow, it remains the case that, especially among the more molecularly oriented of biologists, a muscular embracing of reductionism is expected. Many are reductionists in the sense of the character in Molière's play who had spoken prose all his life without realizing it; it is just the way we are taught to do things and think about them. But there are others who explicitly rejoice in the term, rather than merely implicitly working within its framework.[3] Consider the following episode, dating from 1986, which sets the tone of the issues I need to confront.

The scene is the elegant lecture theatre of London's Royal Society, crammed to overflowing. Many distinguished scientists who had turned up casually only minutes before the proceedings were due to commence found themselves suffering the indignity of being packed into an overflow room to watch the event by video. Karl Popper, the thinking scientist's favourite philosopher, was about to give the first Medawar Lecture. The lecture was named for Popper's lifelong friend and interpreter to the scientific community, Peter Medawar, who, crippled by the series of strokes that were shortly to kill him, sat wheelchair-bound in the front row.

Experimental scientists – at least in the Anglo-Saxon tradition – don't have much time for philosophy and its practitioners; we tend to assume that what we do is obvious and unproblematic, simply holding up to nature as perfect a mirror as can be constructed, and reading the reflection therein. Indeed, younger researchers sometimes speak contemptuously of the 'philosopause' as the age at which their predecessors stopped doing 'serious' work and began thinking aloud instead. Popper, however, for reasons I have already mentioned, was always the exception to this rule, and most of his Royal Society audience knew and admired him mainly through his staunch defence of science against those he – and they – perceived to be its ideological enemies. Unlike other philosophers – and even less like the sociologists of science, viewed with suspicion by many – Popper was regarded as the scientists' friend. But the audience probably guessed from the title of Popper's lecture that defending science was not to be his major theme. Far from it: he was going to have the temerity to challenge one of the core theories of science, that which goes under the name of Darwinism.

The key feature of Darwinism as it is conventionally understood is that it is the external world, the environment, which is constantly setting organisms challenges to their survival. If they meet the challenge, they survive and breed, and their progeny prosper; if they fail, their line diminishes and eventually ceases. It was at this rather passive concept of natural selection that Popper's criticism was aimed. He argued instead for what he called 'active Darwinism', which conceives of the living organism as helping to determine its own fate by itself challenging and modifying its environment to meet its own needs.[4] Staunch evolutionary biologists present at his lecture were not impressed, although this seemingly esoteric distinction is actually by no means anti-Darwinian, whatever some of the great Charles's defenders may claim. Indeed, it is fundamental to our concept of living processes in general, and in particular to what we as humans are and what our destiny might be. However, 'active Darwinism' turned out to be not the only challenge Popper threw at his admirers that evening.

As the allotted hour drew to a close, and the lecturer began to ramble into severe time trouble, he was forced to discard his text and summarize his take-home message via the headings on a handwritten overhead transparency, headed 'Eight reasons why biology cannot be reduced to physics'. Reason number four turned out to be 'because biochemistry cannot be reduced to chemistry'. His conclusion must have been galling to Medawar, who had frequently argued that reductionism is not even second but first nature to scientists.[5] So it was not surprising that, as Popper closed his talk and the chair, President of the Royal Society George Porter (since then elevated, according to bizarre mock-feudal British custom, to the ranks of the peerage), asked for questions, a hand shot up from among those standing at the back of the hall. 'I don't understand why you claim that biochemistry is irreducible to chemistry.' Popper, then in his eighties, was quite deaf, and failed to hear the question until Porter stood up, walked across to him and bellowed into his ear: 'Sir Max Perutz wants to know why you think biochemistry can't be reduced to chemistry.' Popper was never known for his modesty. He stood back, smiled sweetly and said simply: 'Ah yes; I was surprised by that at first. But if you go away and think about it for an evening, you will see that I am right.'

Perutz, himself a Nobel prize-winner for the elucidation of the structure of the oxygen-carrying blood protein haemoglobin, was not amused. His life's work, after all, had been to demonstrate the relevance of chemistry to biology, and he was scarcely used to being brushed off like this, even by a man whose arrogance was legendary. A few weeks later he published a response to Popper's claim.[6] For Perutz, one of the best examples of the fit between chemistry and biochemistry is provided by the way in which the molecular structure of haemoglobin varies between, say, low-altitude, desert-living mammals such as camels, and their cousins the llamas which live at high altitudes in the Andes, where the air is much thinner and the demands on the oxygen-carrying capacity of the blood therefore differ. The molecular structure of camel and llama haemoglobin is in each case subtly modified, the better to fit the conditions in which its owners live. Even different strains of deer mouse adapted to live at different altitudes show genetic differences in the oxygen-carrying capacity of their haemoglobin.[7] Was this not clear evidence that physiology and biochemistry not merely depend on, but are reducible to, the chemistry of the organism's component molecules? Perutz's example matches that given by Steven Weinberg, the opening chapters of whose *Dreams of a Final Theory*[8] pursue the example of why chalk is white down to the atomic level in just this manner.

Game, set and match to Perutz? I think not.[9] But the point of my telling this story is not simply to record the verbal games of a scientific elite or the rather robust style of point-scoring among the philosopher-knights of science. Rather, it is to demonstrate how the reductionism that characterizes the more molecularly oriented biologists is taken for granted, and to start the process of unpicking its many meanings. Indeed, I have already used the term in a number of different ways, without necessarily bothering to stop and clarify which version I have in mind. It is high time I was a little more systematic.

REDUCTIONISM AS METHODOLOGY

First, and perhaps foremost, there is reductionism as a methodology, as discussed in Chapter 2. The living world is characterized by complexity, by flux, by a multitude of interacting processes. We find it easier to understand the phenomena we wish to study if we can hold them relatively isolated from the rest of the world and alter potential variables one at a time: if we can put the marmosets or the chicks into cages, if we can isolate the protein and study its enzymic interactions free from interference with the myriad of other small and large molecules that surround it in living cells. The reasons for doing so are clear. It is hard to make sense of what you observe if several features of a system are changing simultaneously. Reductionist methodology simplifies, and enables one to generate seemingly linear chains of cause and effect.

If I raise the temperature of an enzyme solution by one degree, or alter the acidity of the solution slightly, the catalytic reaction speeds up. I can show what is happening on a simple graph and summarize it in a relatively straightforward equation. Partly because of the way that Western science has developed, to 'capture' a phenomenon mathematically is regarded as one of the supreme scientific achievements: nature tamed and controlled by logic and symbol. But if I make both changes – to acidity and temperature – simultaneously, the two effects are not necessarily additive. Odd things begin to happen. Instead of speeding up, the reaction might even slow down because the combination of increased temperature and acidity makes the delicate protein structure of the enzyme unstable. The equations become complex, or even impossible to formulate. I shall have lost control of the situation, and no longer have the power to predict outcomes. Indeed, until recently the very mathematics has not been available to build models for what might be happening when several variables alter at the same time.

So it is no surprise that reductionist methodology has been so powerful and so attractive over the last three hundred years. It has given us unrivalled insights into the mechanics of the universe, because it often seems to work, at least for relatively simple systems. We can isolate chicks into pens or enzymes into test-tubes and study their reactions. And our experiments are productive, our findings replicable. Within limits, our experiments are successful, our predictions about the world are confirmed. This is why as researchers we get so much pleasure from elegant reductive experiments which give clear-cut conclusions, and why as a teacher I spend much of my time helping my students design such experiments. And, historically, writers and poets who opposed the reduction and mathematization of the universe, the Blakes, or Goethes, the nineteenth-century 'nature philosophers' with their romantic pleas for a non-reductionist alternative, the philosopher Bergson with his vision of a non-physical life force, or their twentieth-century avatars like Sheldrake, have simply been unable to come up with an effective alternative experimental programme.

There is, as will become apparent in later chapters, an alternative, almost underground non-reductionist tradition in biology which stretches back to pre-Darwinian days, to the writings of the French biologists Georges Cuvier and Étienne Geoffroy Saint-Hilaire, through the Bergson-influenced developmental biologist Hans Driesch at the turn of the present century. It had powerful exponents in the 1930s in the form of the Cambridge-based Theoretical Biology Club, whose key figures were Joseph Needham (an embryologist before he became the West's leading expert on the history of Chinese science) and Joseph Woodger,[10] but their voices were and still are drowned out by an almost universal reductionist consensus which insists that, whatever the theoretical critique, reductionism works – or at least, until now it has been made to seem to work. Such simplicity is beguiling.

But living systems are not simple: they involve many interacting variables. Parameters are not fixed; properties are non-linear. And the living world is highly non-uniform. Reductionist methodology is helpful in chemistry, say, because (so far as is known) the chemical world is the same everywhere. In the living world, the exception is nearly always the rule. So if one is not careful, the simplifying constraints that the methodology offers soon cease to be helpful supports to theory, and instead become straitjackets. The Zuckerman trap (see pp. 28–9) awaits us if we are not careful to remember that what happens in the test-tube may be the same, the opposite of, or bear no relationship at all to what happens in the living cell, still less the living organism in its environment. It all depends. And under most circumstances, and until recently, there has been no way of telling – except to try it. This is why, having persuaded my students of the desirability and elegance of reductionist experimental designs, I need to remember, however disagreeable it may seem, that reductionism is not enough when I come to try to interpret my own experiments.

New approaches, it is true, are beginning to make it possible to spring the trap of reductionist methodology. Powerful computers and new mathematical techniques can deal with multiple variables simultaneously. It is at last becoming somewhat easier to model what might be happening if the enzyme is in the cell rather than the test-tube, and to test the predictions from these models by experiment. Of course, testing models is a bit like testing Popper's hypotheses – they are only any good if you can design an experiment based upon them which makes falsifiable predictions. Many models (especially in psychology) remain vacuous because they cannot be meaningfully tested in this way. And at best they are only as good as the data and postulates one feeds in. GIGO – garbage in, garbage out – remains a fact of the computer modeller's life. None the less, the modelling of multivariate systems which can never be approached one variable at a time, such as weather systems, neural processing or three-dimensional protein structures, is becoming increasingly sophisticated and successful. The modellers of such systems rejoice in complexity, and these days take the highly fashionable chaos theory as their route to success. The approach is typified by what has become known as the Santa Fe school, and its prophet is the theoretical biologist Stuart Kauffman, whose recent book *At Home in the Universe* sets out the programme and its claims to be able to handle any problem from the fluctuations of heartbeats on the verge of a heart attack to stock market crashes.[11] In due course such approaches may permit complexity to come into its own. But clear-cut experiments that give unambiguous results, the sort that move effortlessly into the pages of *Nature* or *Science*, and if you are lucky can win you a trip to Stockholm, are for the foreseeable future going to rely on reductionist methodology, even if they do avoid turning method into theory or even ideology.

THEORY REDUCTION

Theory reduction is a term from the philosophical lexicon. One of the aims of science, according to its traditional philosophy, is to simplify, to try to embrace a maximal description of the world within the minimum possible number of laws and variables. The history of science contains a number of examples of what were originally believed to be different phenomena, and were only later discovered to be identical. The classic case is that of the Morning and Evening Stars, regarded as distinct in ancient cosmologies, now understood to be a single entity, the planet Venus, which depending on its motion and position relative to the Earth sometimes appears to 'rise' early in the evening, sometimes to 'set' in the morning. The Morning and Evening Stars are thus both reduced to one object, Venus. Deeper examples come from the development of physics. The sciences of heat and light were once regarded as distinct; today both heat and light are seen as forms of electromagnetic radiation. The separate theories within which each was treated have been reduced to a single unified account.

Such unifications cheer physicists up enormously, to the extent that they sometimes seem obsessed by a reductive drive to simplicity. A major present-day concern is the possibility of developing theories which will embrace all the forces in the universe, strong and weak interactions between subatomic particles, electromagnetic radiation, and so on: so-called Grand Unified Theories or Theories of Everything, sometimes known by their acronyms as GUTs and TOEs. (Nice to see biology getting a look-in at this most abstract of physical levels, if only by virtue of acronyms!) Whether the physical universe can indeed be embraced within such a single theory I have no idea, and while I recognize the goal of simplicity as part of the driving force behind theoretical physicists, I have to confess that what turns them on isn't necessarily what turns on a biologist like myself – but then I also prefer Beethoven to Brahms.

The obsession with simplicity and theory reduction becomes more of a concern to me when it is applied to biology. Some unifications have been immensely powerful, particularly at the interface between biochemistry and chemistry. Stephen Hales, a botanist (today we would call him a plant physiologist) heavily influenced by Newton, gave an elegant theoretical rationale to his reductionist research objectives in the introduction to his classic text *Vegetable Staticks*, published in 1727, from which the epigraph to this chapter comes. But experimentally he was premature. It was not until the end of the eighteenth century that Antoine Lavoisier was able to make the huge conceptual leap which made possible the recognition that the body's 'burning' of the sugar glucose to produce carbon dioxide and water, with the concomitant production of utilizable energy, was in chemical terms the equivalent of oxidation. This understanding – that living processes depended not on some mysterious life force but on chemical reactions which followed the same rules as those of chemistry and could be studied in isolation – led directly to the great reductionist triumphs of the nineteenth and early twentieth century: the elucidation of the basic chemistry of life. More than mere metaphor, homology or analogy, Lavoisier's description of the burning of glucose was an exact description. Perutz must have been very conscious of this when he reacted so strongly to Popper's cavalier dismissal.

Yet there are dangers inherent in such theory reduction. The unification achieved by Lavoisier, and the subsequent demonstration by Friedrich Wöhler in 1828 that an archetypal organic substance, urea, could be synthesized chemically, led to a fully articulated philosophy of mechanical materialism among physiologists. In 1845 four rising German and French physiologists, Hermann von Helmholtz, Karl Ludwig, Emil du Bois-Reymond and Ernst Brücke, swore a famous mutual oath to prove that all bodily processes could be accounted for in physical and chemical terms. Their followers went further, declaring 'Man is what he eats.' The Dutch physiologist Jacob Moleschott put the position most strongly, claiming that 'the brain secretes thoughts like the kidney secretes urine', and that 'genius is a matter of phosphorus'.[12]

But the body's utilization of glucose is, as I shall argue below, not 'just' chemistry. And, even ignoring this nineteenth-century version of sound-bite science, such attempts at theory reduction can lead one into serious errors. For example, textbooks on the philosophy of science offer the reduction of 'gene' to 'DNA' as a parallel case to the identity of the Morning and Evening Stars. But the example is quite inappropriate: 'Morning Star = Evening Star' says simply that, as a result of confusion, two different names were once given to what later turned out to be the same object – like calling a particular animal either a cat or a moggy. However, as will become clear in Chapter 7, 'gene' does not equal 'DNA' in any simple way. 'Gene' and 'DNA' are not (just) two names for the same object. And it is at this point that theory reduction begins to tip over into its much more problematic, full-blown philosophical form.

PHILOSOPHICAL REDUCTIONISM

To underscore the force of philosophical reductionism, let me return to Figure 1.1, the hierarchy of levels in science, on p. 9. The fully fledged philosophical reductionist view of this pyramid is that because science is unitary, and because physics is the most fundamental of the sciences, then an ultimate TOE will be able to reduce chemical theory to a special case of physics, biochemistry to chemistry, physiology to biochemistry, psychology to physiology, and ultimately sociology to psychology – and hence to physics. In its essence, this has been the theoretical claim of molecular biology from its origins in the 1930s. Perutz's limited claim for the collapsing of the biochemistry into the chemistry of haemoglobin is a statement of this position, albeit a relatively modest and specific one. Watson's view that 'there is only one science, physics; everything else is social work' is a characteristically strong version. Linus Pauling sharpened the claim when he advocated 'orthomolecular psychiatry' as a way of resolving mental anguish – it is all a matter of getting the right molecules to the right place in the body. More formal, though no less triumphant, is E. O. Wilson:

> The transition from purely phenomenological to fundamental theory in sociology must await a full neuronal explanation of the human brain ... Cognition will be translated into circuitry ... Having cannibalized psychology, the new neurobiology will yield an enduring set of first principles for sociology.[13]

But what does this conventional diagram of levels really mean? If you look at a University Calendar, you will see that the faculty is divided between departments or schools which are called Psychology, Physiology, Biochemistry, or whatever. Undergraduates study modular courses which have the same neat labels. University libraries contain journals which specialize in each of these subjects, and it is rare to find a physiologist reading a biochemistry journal, still less one devoted to chemistry or physics. Although there are general science journals, like *Nature* in the UK or *Science* in the USA, which report research carried out in many different fields, even the most broadly cultured scientist is likely to be able to understand only one or two of the dozens of articles which appear in their weekly issues.

So what? To go back to Plato, does the division between biochemistry and physiology carve nature at its joints? Or have the two disciplines simply emerged historically because different groups of scientists have chosen to view the world in rather different ways, establishing different languages, different criteria of evidence and proof, and different fields of enquiry – to say nothing of obtaining academic power and prestige *en route* by creating new professorial fiefdoms? There are certainly some who would argue along these lines. Even the official history of Britain's Biochemical Society – the oldest such society in the world – describes the struggles for power as fledgling biochemists sought to liberate themselves from the clutches of the physiology or chemistry departments in which they worked and to establish 'bio-chemistry' as a legitimate discipline in its own right, with independent departments, teaching programmes and professorships.[14] It is not without irony that in the 1960s the biochemists, by then strongly entrenched, fought a similar battle against the recognition of molecular biology, which was once caustically described by nucleic-acid biochemist Erwin Chargaff as 'practicing biochemistry without a licence'.[15]

But biochemistry and physiology are not simply two different university departments whose faculty members may meet in the tea-room, rather as they might also gossip with a literary critic or geographer. Although they speak different languages, use different instruments and read different journals, the phenomena they are studying are the same. It is as if all the university departments corresponding to the levels in Figure 1.1 were stacked on top of one another in a multi-storey building. So what do the different 'levels' represent?

Like many such terms, and like reductionism itself, the way the word 'level' is used in science and philosophy-speak is quite ambiguous. Among its multiple meanings you can find it used to describe simply scale or size, as when people refer to multicellular organisms (measured in metres), organs (measured in centimetres), systems of cells (millimetres), cells (micrometres, or millionths of a metre), and cell membranes (nanometres, or billionths of a metre). It can refer to different body and brain regions (spine, hind brain, mid-brain, forebrain). It can refer to evolution or *phylogeny*, the assumed pathway that leads from single-celled organisms through invertebrates, vertebrates, mammals, primates and humans. And it can refer to development, or *ontogeny*, which begins with genes and ends with complex behaviours. It has a quite different meaning for computer modellers of living processes (algorithmic as opposed to implementation levels); that does not concern me here at all. And finally, to the meanings that do concern me: those to which the formal philosophical terms epistemology and ontology apply. Roughly speaking, *epistemology* refers to how we study and understand the world, *ontology* to our beliefs about how the world 'really' is. So are the levels of the pyramid epistemological? In other words, are they there simply as a consequence of how we choose to work in our different university departments? Or are they ontological, each level corresponding to a different and distinct organization of matter?

Let's go back to something else in Chapter 1: the jumping frog (p. 10). I offered there five types of explanation of the jump, of which the last was the straight reductionist one: the frog jumps because particular muscles in its legs contract sharply, and this contraction occurs because of the biochemical properties of the muscles. Physiology studies the contraction of the muscles, and biochemistry the molecular processes that occur during this contraction. The muscular twitch reported by the physiologist is described by the biochemist in terms of the actin and myosin fragments which comprise the muscle proteins, and whose composition enables them to slide past one another, shortening the muscle (Figure 4.1).

The biochemistry of this process is pretty well understood, down to some of the minutest molecular details. It involves not just the two major proteins, but minor ones too, plus ions such as calcium and magnesium and the ubiquitous 'energy currency' ATP. So why can't we just replace the physiologist's statement about muscle contraction with a statement about actin, myosin, and so on, thus eliminating the need for physiology at all? Of course, there are quite a few biochemists who would raise a cheer at this prospect. But they should beware, because if the physiologist can thus be eliminated, why can't we go on to replace all this talk about actin and myosin with statements about the amino acid sequences of the two proteins, swapping biochemistry for chemistry, just as Perutz was claiming? And doesn't it follow that such chemical chat is much more appropriately cast in terms of the quantum states of the electrons within the molecules? Admittedly, such a statement would grow increasingly cumbersome

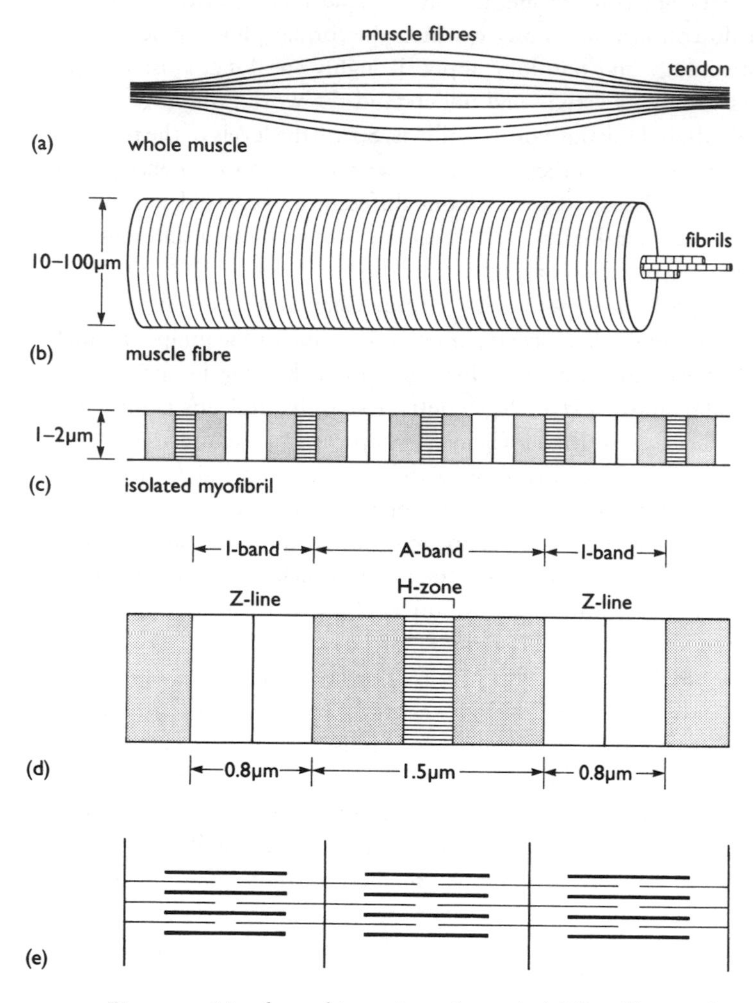

Figure 4.1 Muscle, and its actin and myosin 'sliding filaments'.

as we descended the levels in the pyramid, but we would surely get there in the end. We would successively have eliminated physiology, biochemistry and chemistry in favour of physics. All those university departments could be knocked down, or taken over by the physicists, and undergraduates would only have one subject to study. We would be well on our way to the TOE, and the levels of the pyramid would be entirely an accident of the history of how we study the world; that is, they would be epistemological.

Faced with this consequence, even the most cheerful reductionists begin to jib. Thus Richard Dawkins, in his defence of reductionism against earlier criticisms by myself and others, writes:

> I do not of course subscribe to this ridiculous belief, and I question the good faith of Rose *et al.* in implying that any serious scientist does. The belief attributed to 'reductionists' is exactly equivalent to the following: 'A bus drives fast, because the passengers sitting inside it are all fast runners.' . . . I shall make a distinction between two strategies of reductionist explanation, to be called 'step-by-step' reductionism and 'precipice reductionism'. Precipice reductionists probably do not exist in the world of real scientists, but they have to be mentioned because they are frequently set up as straw men. Step-by-step reductionism is the policy adopted in practice by all scientists with a sincere wish to understand what is going on.[16]

But this analogy studiously misses the point. The belief attributed to reductionists about buses is nothing to do with the passengers the bus is carrying. It is that reductionists wish to explain why the bus drives fast in terms of its mechanical properties, the fact that the engine is turning over fast and it is burning a lot of fuel, and that this in turn is due to the molecular properties of the petrol or diesel and the oxygen with which they interact, which in turn is due to the quantum properties of the atoms of which these molecules are composed. While this is one perfectly appropriate way of describing *how* it is that the bus drives fast, the *why* question relates to the complex framework of public and private transport, schedules, road congestion, driver skills, and so on within which the mechanics of the bus engine are embedded, and these answers to the question cannot be collapsed into either step-by-step or precipice reductionism.

Furthermore, Dawkins' precipice is a slippery place on which to stand. He may wish to descend only a little way down the cliff. Yet having launched himself over the edge, it is extremely hard to see how he can avoid being dashed against the rocks at the bottom. I like the thought of him strolling along the smooth turf of animal behaviour until he arrives at the precipice and then leaping over, carefully donning his hang-gliding wings in advance to avoid descending more than a few metres to the genetic level below. But Watson's 'only atoms' will pull him down, willy-nilly.

In his recent book *Darwin's Dangerous Idea*,[17] the philosopher Daniel Dennett accepts Dawkins' position but, being Dennett, characteristically proposes his own alternative terminology. He is in favour of reductionism, but not *greedy reductionism*. Again, he seems to believe that he can bungee-jump off the cliff edge, but that the elastic will pull him up safely short of the hungry, snapping physicist sharks waiting for him at the bottom. I am sorry, but for all the rhetorical vigour that they expend, I don't see how either the sociobiologist or the philosopher can suspend the laws of gravity and remain in mid-air half way down any precipice. What principle allows them to decide the level at which elimination ceases? Watson's blunt acceptance that there are only atoms is the only logical position – indeed, when he enunciated it in a debate at London's Institute of Contemporary Arts in 1985, the chair, physiologist and Nobel prize-winner Andrew Huxley remonstrated gently: 'Surely Jim, you'll allow cells?' 'No,' Watson replied, 'only atoms.'[18] But then Watson almost certainly sees himself among the sharks rather than those about to be munched. If they pursue this line of argument, Dawkins and Dennett are both doomed to destruction; they merely have the right to choose between death on the rocks and dinner for sharks.

Leave aside for the moment the fact that such reductions, even if theoretically possible, are currently beyond the wildest dreams of physics, which cannot yet solve the problems of three simultaneously interacting particles, nor, I am told, predict the properties of water from knowledge of the quantum states of the oxygen and hydrogen atoms that comprise it. Instead, let us try to get clear what reductive elimination is about. Are we trying to describe a *causal* relationship

– that the biochemistry is causally *responsible* for the physiological event? If so, this is a very different use of the word cause from the way in which we normally employ it to describe the relationship in time between cause and effect – one event necessarily and specifically following from another.

In the common-sense understanding of cause preceding effect, the proximal cause of the muscle twitch is provided by the physiological description of impulses travelling from the frog's brain down the motor nerves to its muscles. As pointed out on p. 13, the sliding of the actin and myosin filaments does not *precede* the muscle contraction; in an important sense it *is* the muscle contraction – or at least part of it. It only confuses matters to use the word 'cause' to describe both a temporal cause-and-effect sequence and also this special relationship between the physiological and the biochemical descriptions of the processes. We are really making not a causal but an *identity* statement here. (I am not suggesting that we revert to Aristotle's terminology of material, formal, efficient and final causes, rather that it would be helpful to restrict the term 'cause' to temporally defined, within-level relationships.)

The most straightforward way of describing the relationship between the physiological and biochemical statements about the muscle twitch would be to refer to them as if one were simply dealing with two different languages, and translating between them. You can say 'cat' in English and '*gatto*' in Italian, and you will be talking about the same four-legged, furry purring object. No one assumes that the task of a translator is to eliminate *gatto* in favour of cat – or vice versa – that one of the two languages is the 'real' way to talk about the cat/*gatto* object and the other is used only because we don't yet understand the real nature of the beast. Why can't we then simply say that the sliding actin and myosin filaments are the biochemist's way of talking about what the physiologist calls a muscle twitch?

Where does this leave the physiologist with regard to the temporal relations of cause and effect in which the muscle twitch is preceded by a signal travelling down the motor nerve? Biochemists can also describe in exquisite detail the processes that go on when such a signal (an *action potential*) passes down a nerve fibre. The potential depends

(a)

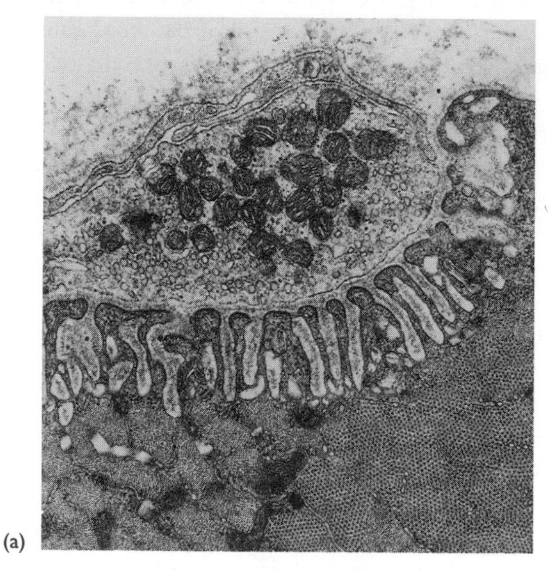

(b)

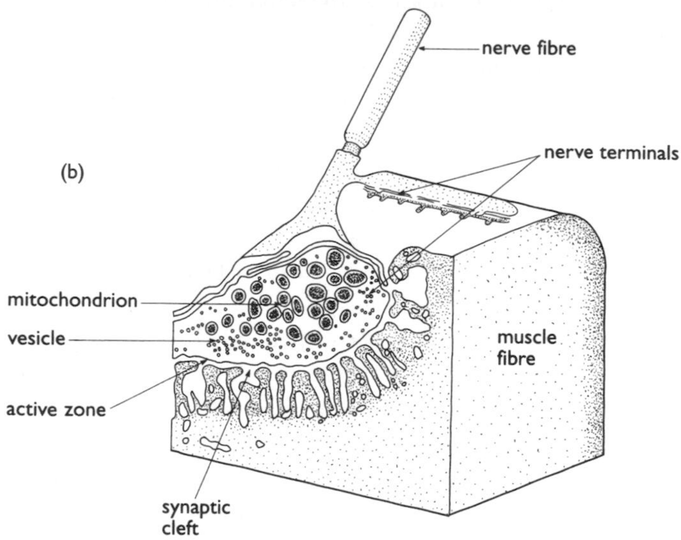

Figure 4.2 The neuromuscular junction: (a) electron micrograph and (b) diagram.

on the lipid and protein structure of the nerve cell membrane, the differential distribution across the membrane of the ions of potassium and sodium, the ubiquitous ATP, and so on. At the junction between nerve and muscle (the *synapse*), defined in considerable anatomical detail, there are tiny membrane-bound packages (*vesicles*) containing transmitter molecules which can be released when the action potential arrives at the synapse, diffuse across to the muscle, and there begin the biochemical cascade that ultimately results in the actin and myosin filaments sliding past one another. The process is summed up in Figure 4.2.

Notice that the description that I have given of the biochemical mechanisms now enables me to produce a temporal cause-and-effect sequence in biochemical language too. So I can match the two sequences, described in the two languages of physiologese and biochemese, as in Figure 4.3. But isn't an identity statement of this sort exactly what reductionist philosophy is demanding? Despite all my talk of translation, haven't I simply made the eliminative step? The world is an ontological unity, and the apparent epistemological diversity is trivial. Well, no; but then I assume you would expect me to say that. So let me spell out why I want to insist that the first of those statements – the claim for ontological unity – is valid, even though the second – the denial of the significance of epistemological diversity – remains false.

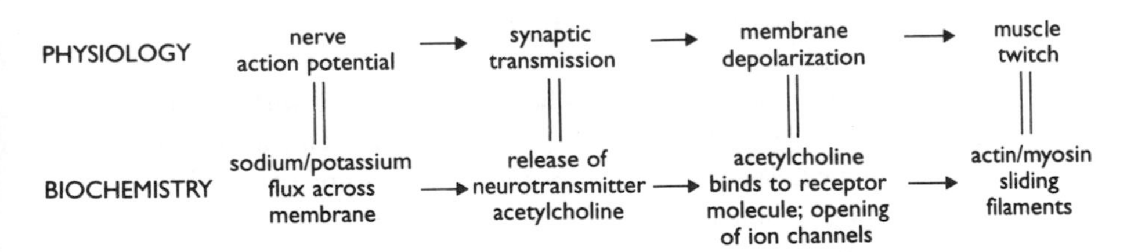

Figure 4.3 A muscle 'twitch' in two languages.

LEVELS REVISITED

In order for reductionism to be valid, we have first to make a further assumption, and second to ignore several other central features of the example. The additional assumption is the premise built into Figure 1.1, that the different 'levels' of the pyramid are really arranged hierarchically, so that the lower they are the more 'fundamental' they are. Because physiology is positioned above biochemistry, it, not biochemistry, is the one that should be eliminated, and so on ultimately down to physics, which is the ground floor of our multi-storey university science block. What are the grounds for providing this ordering, for defining 'fundamental' in this way?

Two arguments are sometimes given. The first is that the lower levels represent more generally applicable principles. For instance, the findings ('laws') of atomic physics are believed to apply throughout the entire universe, while those of physiology relate, so far as is known, only to the special case of living systems here on Earth (though presumably they would apply elsewhere too, given the right conditions). The second is that the higher levels represent more complex states of matter than the lower. Matter takes on a more organized form in a fish than in the water through which the fish swims. From these two premises it is said to follow that, while all physical facts, principles and laws apply to the living systems that the physiologists study, the reverse is not true. Physiological principles are not applicable to stones or planets.

There is a counter-argument. The inanimate world with its 'laws' of physics and chemistry is only knowable, the laws only articulated, because of the existence of matter organized in the degree of complexity that is found in human brains and societies. Without the social and cerebral activities of scientists, the science of physics simply wouldn't exist. Perhaps Figure 1.1 should cease to be drawn as a pyramid at all, but instead should loop back on itself, like a doughnut, with sociology and psychology forming the basement of the building. But even if we discard the pyramid with its levels and implied directionality, there are deeper problems with reducibility.

The first arises out of a refinement of the discussion about the translation between physiology, biochemistry and chemistry. To return to the case of the jumping frog: the biochemistry of the muscle twitch, the synaptic transmission or the action potential, do not occur in the isolation of a test-tube. Muscle fibres, synapses and nerves are all anatomical structures, each with specific locations within the frog. The translation from physiology to biochemistry and chemistry is incomplete without reference to this anatomy, which means that the biochemistry of the action potential occurs in a different place and at a different time from that of muscle contraction. The relationships between these different molecular processes are organized in space and time in a manner which is not implicit in their chemistry. As long ago as the 1930s, it was shown that actin and myosin fibres can be reconstituted as filaments in a test-tube – one of the first known examples of the self-organizing properties of proteins, described in the chapters that follow. In the presence of ATP they will shorten just as they would during muscular contraction. But they do not thereby *become* a contracting muscle fibre. This requires a set of irreducible organizing relations, implicit in the physiology of the process but absent from either the biochemistry or chemistry, which define the *functions* of the muscle. This, I believe, was what Popper meant by the slogan that so outraged Perutz.

It is also, I suppose, what might be conveyed by the well-known phrase that 'the whole is more than the sum of its parts', and I might be happy with it had the slogan not become surrounded by a sort of mystic aura. But what can be asserted, without retreating into hand-waving or New Age sloganeering, is that the key feature distinguishing a lower 'level' of the pyramid from those above it is that at each level new interactions and relationships appear between the component parts – relationships which cannot be inferred simply by taking the system to pieces. Furthermore, the claim makes an additional important assertion. Philosophical reductionism implies that, whatever higher-order properties emerge and however they do so, they are always somehow secondary to lower-order ones. The lower the order, the greater the primacy. Parts come before wholes. Yet, whatever the case may be for the properties of physical and chemical systems, the nature of evolutionary and developmental processes in biology means that there is no such necessary primacy. Wholes, emerging, may in themselves constrain or demand the appearance of parts. Arthur Koestler, arguing in the 1970s for *Beyond Reductionism*,[19] described each 'level' as having a Janus-like relationship to the others. To that immediately below it, it was unitary (or, as he named it, a holon[20]), while to that above it, it was an assemblage of components. The ontological unity of the universe then consists not of a pyramid of levels, but of a nested hierarchy of holons, and might be drawn as in Figure 4.4.

Let me put it less abstractly. Our younger son Ben in his youth once tried to get the sewing machine to work, and when he failed he began to take it apart to try to discover the source of his problem. When he came to reassemble it, he found some parts inexplicably left over. He couldn't see where they fitted in, so he left them in a neat heap by the side. They were spare, he said, not needed. His problem

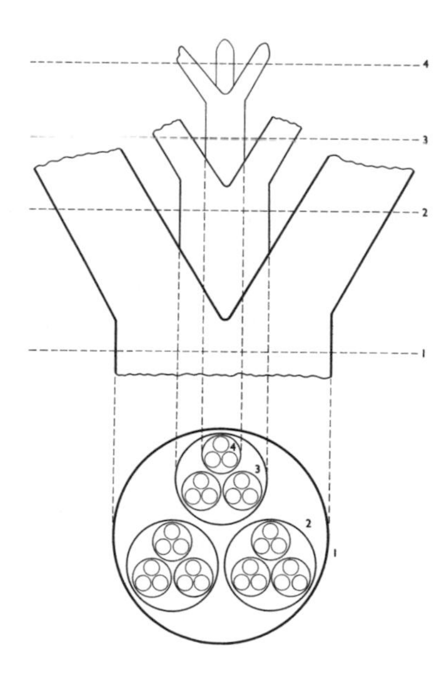

Figure 4.4 Nested holons, as envisaged by Arthur Koestler.

was that of the typical reductionist trying to assemble physiology from biochemistry. Unless you know the function of the parts in the system, you can't understand what they are for or how they fit together. I have a similar problem with car engines, even though I know what the engine is supposed to do, just as Ben knew what the sewing machine was for.

But to give a group of Martian visitors the bits of a car engine or sewing machine and ask them to assemble it without any knowledge of its function or purpose would be to present them with an impossible conundrum. Neither machine is explicable without the knowledge that in our society clothes and curtains are stitched together out of fabric, and that we Earthlings move from one locale to another by means of individual person-carriers operating within transportation systems involving not merely internal combustion engines, but also roads and appropriately spaced networks of petrol stations, starting-points and destinations. That is, to understand any piece of machinery you need to know not merely its composition but its role in the larger system of which it is a part. This is why the jumping frog requires not merely the within-level, temporally causal type of explanation and the reductionist actin-and-myosin type explanation, but also the 'top-down' or system-level explanation. If you don't know that the frog has just seen a snake and is trying not to become its dinner, you understand only a fraction of what is going on. A living organism cannot exist independently from its environment, with its constant interchange of energy and information, threats and promises. Living systems are by definition open ones. Our multi-storey science block is part of a university. Without language, history and geography departments, science would be meaningless.

Our world may be – is, I would claim – an ontological unity, but to understand it we need the epistemological diversity that the different levels of explanation offer. And if you still aren't convinced, and believe you can hang-glide off Dawkins' precipice without coming to harm, why bother reading the words, paragraphs and chapters of which this book is composed? All you need to do is examine the individual letters on the page, call in an analytical chemist to give you the formula of the printer's ink, and a microscopist to describe the fibre structures of which the paper is composed. This is why reductionism, once it ceases to be merely methodological, when experimenters can just about hang on to the edge of the precipice by their fingernails, so rapidly tumbles into ideology.

NOTES

1. Lynda Birke, *Feminism, Animals and Science*.
2. See E. O. Wilson and S. Kellert (eds), *Biophilia*; for a contrasting view as to the robustness of nature, see the chapter 'God, Gaia and Biophilia' by Dorion Sagan and Lynn Margulis in that book, and also the critique by Stephen Budiansky, *Nature's Keepers*.
3. They include, to name but a few, sociobiologists such as E. O. Wilson, molecular biologists such as Francis Crick and James Watson, neuroscientists such as Jean-Pierre Changeux, and theoretician-polemicists such as Richard Dawkins.
4. From Popper's 1st Medawar Lecture, delivered to The Royal Society in 1986. As yet it is unpublished, though an audio cassette is available from the Society's library; publication from Popper's notes is planned.
5. Peter Medawar, 'A view from the left'.
6. Max Perutz, 'A new view of Darwinism'.
7. *Ibid*.
8. Steven Weinberg, *Dreams of a Final Theory*.
9. See my own reply to Perutz: Steven Rose, 'Reflections on reductionism', and ensuing correspondence in *Trends in Biochemical Science*: Perutz's disclaimer is published as 'Reply, from Perutz, on reductionism', and there followed 'Steven Rose replies'.
10. Donna Haraway, *Crystals, Fabrics and Fields*. Recent attempts to recapture this tradition have been made by the theoretical biologists Gerry Webster and Brian Goodwin, *Form and Transformation*.
11. Stuart Kauffman, *At Home in the Universe*.
12. From Moleschott's 1852 text *Das Kreislauf des Lebens*, quoted in Donald Fleming's introduction to Jacques Loeb's *The Mechanistic Conception of Life*.
13. Edward O. Wilson, *Sociobiology: The New Synthesis*.
14. Trevor W. Goodwin, *A History of the Biochemical Society 1911–1986*.
15. For Chargaff's scepticism about the new molecular biology, see Robert Olby, *The Path to the Double Helix*.

16. Richard Dawkins, 'Sociobiology: The new storm in a teacup'.
17. Daniel C. Dennett, *Darwin's Dangerous Idea.*
18. James Watson, 'Biology: A necessarily limitless vista'.
19. Arthur Koestler and J. R. Smythies, *Beyond Reductionism.*
20. The developmental biologist Antonio Garcia-Bellido refers to Koestler's holons as *nodules* in 'How organisms are put together'.

5

Genes and Organisms

. . . once 'information' has passed into the protein it cannot get out again. Francis Crick, 'On protein synthesis'

GENES AND GENETICS

The trajectory of any organism through time and space – its personal lifeline – is unique. Although each individual resembles all others of the same species, and resembles more closely still its parents and siblings, no two are exactly the same – those who know them well enough can tell even identical human twins apart. What confers these similarities, these identities and differences, on the space time trajectories of life? Such core questions have now occupied biologists for more than a century. They are the objects of study of two different biological disciplines, genetics and developmental biology, which began by asking rather similar questions about the nature of life, but at a key point in their history became damagingly separated one from the other. This has resulted in conceptual confusions which have persisted well into the present-day era of high-tech molecular biology. To appreciate the consequences of these confusions, we have to go back into the history of genetic and developmental thinking. Biology's own history is centrally engaged within these current disputes. The past, as so often, is the key to the present, and the current disputes within biology can be properly understood only by reference to the history of the study of living processes.

While questions about the origins and development of living creatures had concerned biologists long before the term 'biology' itself was introduced, in 1802, the basis of our current understanding is the work of Gregor Mendel, who began his famous experiments on the colour and shape of successive generations of pea seeds in the garden of his abbey in Brno, in what is now the Czech Republic, publishing them in 1865. He not only showed that these two features (yellow versus green; round versus wrinkled) were transmitted between generations independently of each other, but he introduced what was, for the experimental biology of the period, a quite novel approach. Unlike his predecessors, though contemporaneously with others such as the polymath eugenicist Francis Galton in England, Mendel did more than just describe his findings qualitatively: he *counted*. He observed that, depending on the nature of the parental plants which he crossed, the features green and yellow, wrinkled and round, appeared in successive generations in simple and reproducible ratios.

He started from what became known as 'pure lines' of peas, which had been kept separate and bred true for generations. He then crossed two lines. For instance, if he fertilized a green-pea plant with pollen from a yellow-pea plant, all the offspring had yellow peas. However, if he now crossed these offspring among themselves, some of the plants that resulted bore yellow peas, and some bore green peas. Thus inheritance was discrete – the yellow and green did not mingle to produce some intermediate colour. Furthermore, the capacity to produce green peas was not lost in the original green–yellow cross, it was merely masked. And finally, the second generation always produced green and yellow peas in an almost precise ratio, one green to three yellow. What was true of the colours of the peas was also true of the other characteristics, such as round versus wrinkled, that he studied.

It was as if each observable feature of the pea plant, each of its surface properties, or *characters*, was represented within the plant by some unobservable particle or (in modern terms) store of information, on the basis of which the colour and shape of the succeeding generation of pea seeds was determined; indeed, these mysterious factors became known as *determinants*. Each offspring of a mating received a pair of determinants, one from each parent. If both members of a pair were, say, green (that is, if the pair was *homozygous*), the offspring would be green; if both were yellow, then the offspring would be yellow. But if the inheritance was one green, one yellow determinant (that is, if the pair was *heterozygous*), then the offspring was yellow. The yellow determinant was *dominant* over the green, and the green was said to be *recessive*. The famous three-to-one ratio follows very simply from these assumptions. If the determinants are represented as Y and G, and each plant of the original pure lines has two copies of its determinant (YY or GG), and the first cross, which is yellow, inherits one determinant from each of its parents, then it must be YG. If two YG plants are now crossed, the possible combinations in the offspring are one each of YY and GG, and two YGs. And as Y is dominant, all plants with a Y determinant have yellow peas and only the GG plant will have green peas. Three to one – simple (Figure 5.1).

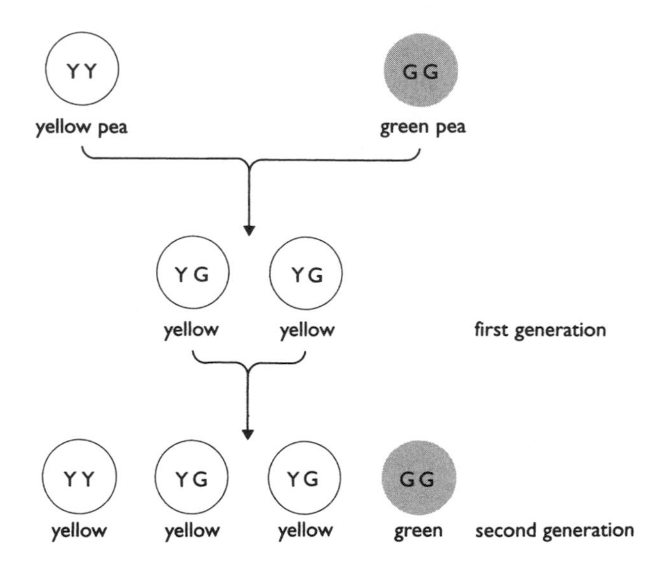

Figure 5.1 Mendel's ratios for yellow and green peas.

Sexual reproduction in plants or animals differs fundamentally from the asexual reproduction or budding off which is the main (though not the only) way of reproduction in bacteria and other single-celled organisms. In asexual reproduction the two daughter cells produced by the budding have only one parent, and can therefore receive only one type of determinant; the mixing of determinants that happens in sexual reproduction is impossible. Daughter cells are all identical to their parent cells – *clones*, therefore. It turns out, incidentally, that sexual mixing has certain genetic and evolutionary benefits quite apart from the pleasure it provides the sexual partners, which is presumably why asexually reproducing organisms like bacteria, in whom it is difficult to detect anything observably akin to human pleasure, occasionally go in for bouts of sex (called in their case conjugation) in which genetic material is transferred from one, regarded for these purposes as male, to another, regarded as female. On the other hand, as any gardener knows, many plants have an a.c./ d.c. approach to the whole reproductive business.

Like all good experimenters, Mendel was lucky.[1] The characters he studied seemed discrete: there was no intermediate state between being wrinkled or round, yellow or green. By contrast, the characters that interested Galton – human features such as height, or strength of hand-grip, or head circumference or intelligence – were not discrete but varied continuously across a broad range. Furthermore, the offspring of people of disparate heights, rather than following either one or other parent in a Mendelian manner, tended to occupy some middle territory between the two. Such continuously varying characteristics seemed, in Galton's studies, to blend – an observation which haunted Darwin, and which, as will become clearer in Chapter 7, resulted in endless problems for his model of natural selection as the motor of evolution. This isn't just a difference between plants and animals. One reason why Mendel's work was ignored for some forty years was that when the doyen of contemporary European botanists, Karl Wilhelm von Nägeli, with whom Mendel corresponded, suggested he repeat his studies with a different plant species, the experiments failed to show clear-cut transmission ratios.[2] Nägeli himself was sceptical about Mendel's theories, but the real sticking-point is that the ratios only appear in particular instances.

Mendel's results were independently rediscovered in 1900 by researchers in Tübingen, Vienna and Amsterdam[3] who between them founded the modern science of what became known, following another early Mendelian enthusiast, William Bateson, in Cambridge, as *genetics*. Mendel's ratios ceased to be a special property of peas when they were shown to describe the transmission of discrete characters in many other species. The study of family trees running back through three or more generations of humans (so-called *pedigrees*) showed that, in our species too, certain clear-cut features such as eye colour or the ability to roll one's tongue were inherited in proportions that could be fitted to Mendelian ratios.

New terms appeared: the individual hidden determinants of surface characters became *genes*, and the total of an individual's genes formed its *genotype*. The specific copies of any given gene possessed by an individual (for instance, the Y and G determinants in peas) were termed *alleles*. The surface characters themselves comprised the individual's *phenotype*. It is important to recognize that none of these terms was very precisely defined, and almost from their introduction they meant different things to different researchers, varying from the specific features of any individual of a species to some Platonically idealized 'species-type' to which all actually existing members of the species more or less approximated. Indeed, more recently the term 'genotype' has tended to be dropped because it carries on its shoulders precisely such Platonic baggage; these days the sum of an individual's genes is more usually referred to as its *genome*. An element of Platonism was there from the beginning, though, and remains today in some quarters. Genes were 'essences', hard impenetrable units which might even be seen as counterparts of the atoms of the early-twentieth-century physicists: the ultimate, indivisible units on which outward forms depended; the unmoved movers, unchanged changers, within each organism.

'Phenotype' is similarly ambiguous, and is used to refer to any or all observable or measurable features of an organism, from the presence of a particular enzyme to hair colour or body feature, or even a piece of characteristic behaviour such as the gait of a walker. In his book *The Extended Phenotype*,[4] Dawkins even goes so far as to describe aspects of the external environment of an organism as part of its phenotype – for instance, he sees the dam that a beaver constructs as part of that beaver's phenotype. This idiosyncratic extension of the term simply makes the problem of terminology even harder, for the dam is the product not of the activity of a single individual, but of the collective labours of many beavers. It also harbours a multitude of insect species which enjoy the special features its environment provides. If the dam is a phenotype, it is the phenotype of a community, not of an individual, and its relationship to any individual's genes, genotype or genome is thus distinctly tenuous – an issue to which I shall return in Chapter 8.

Following their rediscovery, Mendel's ratios were to dominate thinking in the infant science of genetics for several decades, just as they still form the starting-point for most school biology texts. However, the distinction between Mendelian discontinuous variation and Galtonian continuous variation remained problematic through the 1910s and 1920s.[5] The Galtonian tradition was carried forward by Galton's pupil, protégé and successor, Karl Pearson, in London. Pearson was a formidable mathematician, and because the complex data derived from the characters (or *traits*) that he measured failed to yield neat either/or, green/yellow divisions, he set about developing many of the statistical methods still in use today to analyse complex data (indeed, the histories of genetics and of statistics have been interlocked ever since). The resolution of the conflict between Mendelians and Galtonians came to depend on the recognition that continuous variation in features such as height could be interpreted as a consequence of the interaction of many genes, each with a small effect on the final outcome.

But as time wore on, the number of observed divergences from simple Mendelian ratios steadily increased. An early finding was that some characters are expressed only in one sex. Colour-blindness or haemophilia, for instance, occur only in males, although both are capable of being inherited through the female line. Haemophilia is notorious in a famous pedigree, traceable from Queen Victoria, among the intermarrying royal families of Europe and culminating in the son and heir of the Romanov czars of Russia, executed following the Bolshevik Revolution of 1917. Such characters are said to be *sex-linked*. Other divergences from the ratios are less straightforward, and the models developed to account for them became more and more complex. However complicated and varied the observed phenotypes, the modellers were still determined to explain them on the basis of the interaction of the indivisible causal particles they conceived genes to be. If the ratios didn't work, other factors had to be obscuring the proper functioning of the genes, just as the Devil can interfere with God's purposes. Genes were said to be *partially dominant*, or to show *incomplete penetrance*.

Indeed, once these possibilities are admitted, there is virtually no distribution of phenotypes found in the population to which a genetic model cannot be fitted. In the traditional Popperian sense, such genetic models, which may become as complex as the 'wheels within wheels' invoked by pre-Copernican astronomy to account for the motion of the planets, are strictly unfalsifiable. Given enough assumptions, any model can be 'fixed'. The ease with which this can be done was brought home to me a few years ago in the course of a conversation with an eminent behavioural geneticist. We were discussing the genetics of schizophrenia, and I described some recent evidence that the incidence of the diagnosis in Britain is much higher in the children of black/white relationships than in either of the parental populations, the indigenous whites or the black immigrants from the Caribbean. No simple genetic model could fit this data, and one obvious interpretation is that the schizophrenia diagnosis is a consequence of the strains of growing up as the child of a mixed relationship in a racist society. It took the geneticist scarcely a moment to generate the alternative genetic model – assortative mating. That is, you have already to be mad to consider mating with a person of a different colour!

By the 1920s, a number of genetically transmitted diseases had been identified in humans (described as 'inborn errors of metabolism' as early as 1909 by one of the founders of the field, the medical doctor Archibald Garrod). Some at least – notably blood disorders such as sickle-cell anaemia – appeared to be inherited in Mendelian or quasi-Mendelian fashion. By this time the eugenics movement, which Galton had founded, was arguing that everything from feeble-mindedness to sexual promiscuity and criminality were also heritable in this way – intertwining now not merely genetics with statistics, but both with psychometry and eugenics.[6] Thus began the long journey which led, through the sterilization acts and anti-immigration legislation in the USA, to the Nazi death camps and beyond. This history, which has been recounted many painful times,[7] forms part of the tortured public legacy of modern genetics and cannot entirely be transcended, for it still colours reductionist thinking in biology. This legacy makes genetics, along with nuclear physics, perhaps the two areas of science of greatest general public concern, but it is not my intention to pursue that theme further here. Instead I want to focus again on issues of lifelines, of the trajectories of individuals in time and space.

DEVELOPMENT

While the Mendelian rediscoverers were busy defining the phenotypic features they observed as the products of hypothesized genes, other biologists were looking at organisms from quite a different perspective. How, they asked, does the union of egg and sperm ultimately produce an organism which may consist of a hundred trillion (10^{14}) such cells, differentiated into tissues and organs, precisely located in space in relationship to one another?

The problem was first tackled by observation. An optical microscope, and an animal whose fertilized eggs could readily be observed during development – sea urchins and amphibia such as frogs became the favoured objects of study – were all that was required. The fertilized cells could be seen to divide within about an hour; one cell became two, two became four, four became eight . . . This differs fundamentally from the division that occurs at fertilization, in which the fertilized offspring cell receives different sets of Mendelian determinants or genes, one allele from each parent. Instead, each daughter cell resulting from cell division (*mitosis*) receives an exact copy of the genes present in its parent, as in asexual reproduction.

Within about eight hours the dividing cells have formed a hollow ball, or *blastula*, one cell thick and containing about a thousand cells in all (Figure 5.2). Then the cell ball begins to change shape, as if it were being pushed in at one point until the indentation reaches the

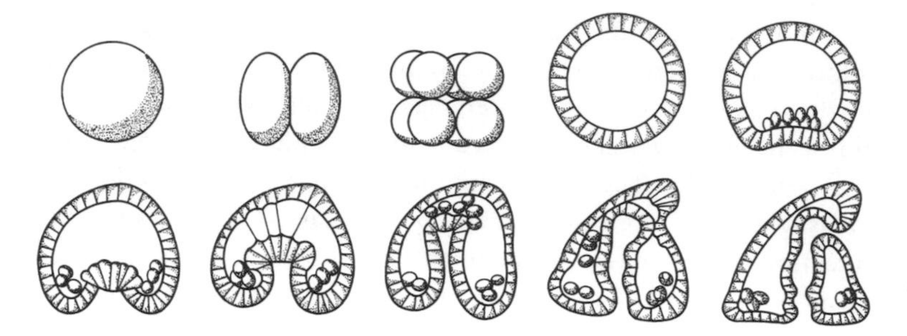

Figure 5.2 The early stages of cell division: from egg to blastula and gastrula.

inside of the opposite wall of cells. This is called *gastrulation*. As division proceeds, the gastrula twists and turns, and develops further indentations, regions become pinched off entirely to form independent structures. After a surprisingly small number of cell divisions (after all, it takes only twenty divisions for a single cell to multiply to over a million) the cell mass is recognizable as a miniature version of the adult – or in the case of the frog, of its tadpole stage of life.[8]

Microscopic examination of dividing cells revealed something more, even with the limited methods available by the end of the nineteenth century. As a eukaryotic cell (one with a nucleus) prepares to divide, previously invisible thread-like structures begin to coalesce and appear within its nucleus. These structures can take up the dyes that the microscopists were then using, and therefore appeared coloured; in recognition of this property, they were called *chromosomes*. (In prokaryotes like bacteria, which have no nuclei, the chromosomes appear in the cytoplasm.) The chromosomes of any organism have characteristic shapes, and in the microscope they look like small twisted ribbons, each bearing a specific pattern of horizontal stripes or bands, a bit like an irregular ladder. In each cell (except the sex cells or *gametes* – eggs and sperm) chromosomes exist in matched pairs. As mitosis (division) proceeds, each chromosome appears to double, as if making a copy of itself. The copies begin to separate, moving to different parts of the nucleus. The nucleus itself then becomes pinched in the middle and splits in two, and finally so does the entire cell, so where there was one, there are now two daughters, each bearing a full set of the chromosomes of their parent. The sequence is shown in Figure 5.3.

This precise internal cellular dance of the chromosomes, and the rhythm of cell division unrolling in a seamless sequence, was and still is fascinating to observe, but it operates according to rules which the early embryologists found hard to fathom. Indeed, they saw the process as encapsulating everything that distinguishes life from non-life. For some, the only explanation was that the developing embryo was imbued by an *élan vital*, a life force irreducible to mere mechanism. To most, however, this conclusion was unacceptable: they were observing a complex piece of living clockwork, which if it could be taken to pieces would reveal its works. Whichever philosophy one adopted, the dividing ball of cells was splendidly accessible to experimental manipulation. What would happen, for instance, if one removed a portion of the dividing cell ball, or cut it neatly in half? The results confused researchers for decades, for the conclusion seemed to be 'it all depends': depends on the organism; depends on how many divisions the ball of cells has made prior to the cut; depends from where in the ball one removes the sample.

Thus in the 1880s, one of the founders of developmental biology, Wilhelm Roux, killed (with a hot needle) one of the two daughter cells resulting from the first division of a frog's egg. The result, in accord with his mechanistic beliefs, was that the surviving cell gave rise to only half an embryo. Embryological development was thus the mechanical unfolding of determinate stages, with irreversible differentiation of function between each cell. By contrast, his pupil Hans Driesch announced in 1891 that if he performed the same experiment with sea-urchin eggs at the two- or four-cell stage, he obtained perfectly formed adults, but each just one-half or one-quarter the normal size. For Driesch this seemed a complete refutation of the mechanistic view of life – after all, if a machine is taken apart the individual pieces can never be turned into two or more complete functioning machines of the original type. As a result he espoused a modern version of vitalism ('entelechy'), a power which inhabits all living cells and ensures their harmonious functioning.[9]

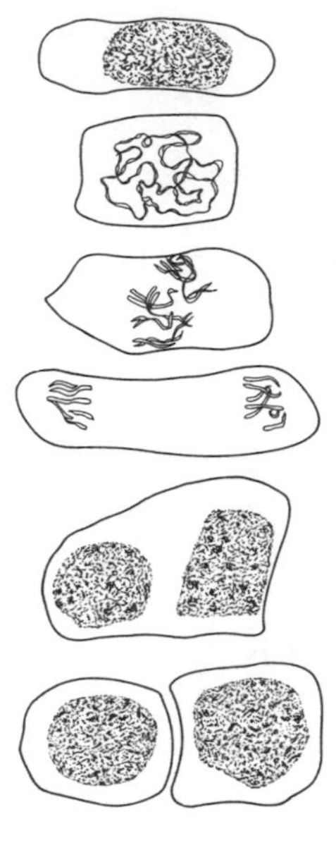

Figure 5.3 A cell, its nucleus and chromosomes in cell division. Drawn from a photograph of cells from the tip of a root of a crocus.

Driesch's mystic formulations, oddly analogous in his day to Sheldrake's in ours, were in turn countered by another of Roux's followers, Jacques Loeb, who reinterpreted the experiments to demonstrate that indeed the outcome all depends. Under certain circumstances each half of the severed cell ball will grow into a small but perfectly formed adult. Under others, one portion will develop perfectly and the second will not be viable. Under still others, an imperfect organism develops, lacking some vital function or body part. The results could be systematized as follows: depending on the organism, at early stages in the cell division process each cell still retains all the determinants – the information or the genes, call them what you will – to make an entire offspring; at later stages some regions of the developing ball of cells retain this capacity but others do not; later still the capacity is entirely lost, and the developmental fate of each region of the cell ball is fixed and cannot be modulated. (At least this is true of animals. In plants such a fate is not inevitable: in appropriate circumstances even a tiny sliver of adult carrot plant can be induced to grow into a new, fully formed carrot – a clone of the original, therefore.)

Loeb went on to formulate his own grand theory, which he called *The Mechanistic Conception of Life* in a book – or rather a manifesto – published in 1912. In it he claimed that organisms *are* machines. Behaviour, even of the most complex kind, can be broken down into a series of mechanical *tropisms*, tendencies to move towards or away from the light, or in response to gravity, or whatever. Furthermore, simple biochemical mechanisms could account for such tropisms. If Driesch was the Sheldrake of his day, Loeb was certainly the Dawkins. Consider this statement from Dawkins:[10]

> A bat is a machine, whose internal electronics are so wired up that its wing muscles cause it to home in on insects, as an unconscious guided missile homes in on an aeroplane.

Loeb could not have put it better, though his mechanistic metaphors were limited by the power of 1912 technology. Even if such formulations now seem simplistic, their organizing, ideological power was enormous. Loeb worked at the Rockefeller Institute (now university) in New York, and his thinking helped shape the Rockefeller's programme of funding of molecular biology which dominated the field from the 1920s to the 1950s.[11]

To resolve these paradoxes, later generations of embryologists embarked on more complex experiments. Parts of the developing embryo were transplanted to other regions. And again, the results depended on the exact conditions. Thus in 1924 Hans Spemann and Hilde Mangold grafted a piece of tissue from a region of the newt gastrula to the opposite side of another embryo, and observed a whole second embryo developing in the region of the graft. The graft had changed the fate of the cells around it, *inducing* the formation of the second embryo; they called the graft the *organizer*. (In 1935, not long after Mangold's tragic accidental death, Spemann was awarded a Nobel prize.) But nothing is simple in developmental biology. Sometimes the fate of a transplant is determined by the environment into which it has been transplanted; sometimes it carries its own fate with it. Transplant a group of cells from the region of a developing insect destined to become a leg, and insert it into the head region. Depending on the age of the embryo, and hence on the number of divisions it has undergone since fertilization, the transplanted tissue may be incorporated into the developing head, or it may develop into an additional leg projecting anomalously from the head.

Mendel had imagined that each organism contained, as it were, a bag of separate determinants, and that during development different determinants were dispatched to different regions. But what happens during mitosis turns out to be different: each cell receives an identical set of determinants or genes. In the early stages of development, therefore, any of the newborn cells can divide and produce an entire organism; it is said to be *totipotent*. Later, however, although all the genes are still present in all the cells, which genes are active (*expressed* is the term used) depends on the developmental history of the particular cell: that is, on how many times the cell line that leads to it has divided, and whereabouts in the developing embryo the cell is located. Thus gene expression varies in both time and space.

GENES AND CHROMOSOMES

Genes, meanwhile, remained abstract, invisible determinants. Mendel's laws of transmission and the independent segregation of genes were confirmed and extended during the 1920s and 1930s when Thomas Hunt Morgan and his team, first in Columbia, New York, and later at Caltech, found a suitable animal model for their study – the ubiquitous and rapidly breeding insect commonly known as the fruit fly or vinegar fly, *Drosophila melanogaster*. These are the tiny black specks which collect in large numbers on any piece of ripe or rotting fruit. *Drosophila* breed very rapidly, and among any large population of them Morgan found some which seemed unusual – for instance, they had white rather than red eyes, or a different pattern of veins on their wings. These unusual characters are transmitted in a Mendelian manner. His colleague Hermann Muller showed that the proportion of unusual characters could be greatly increased by stressing the flies in some way – for instance by exposing them to not-quite toxic concentrations of particular chemicals, or to radiation such as X-rays. The strange features of the flies that result are *mutations*, and could be studied in both the mutated individuals and their offspring.

Morgan had begun his research life not as a Mendelian geneticist, but as an embryologist, and the other reason why *Drosophila* were interesting to him was that their cells contained unusually large and readily visible chromosomes.[12] He could examine the structure and appearance of the chromosomes during cell division and compare them in the mutated and the normal (or 'wild-type', as they came to be known) populations. This enabled him to take the next key step in the history of genetics. The abstract determinants called genes, it turned out, had a physical location in the cell. Genes were on chromosomes, and were thus distributed to daughter cells during mitosis by the division of the chromosomes between them. During sex, the fertilized egg received its genes, half from each parent, in the form of the single set of chromosomes that each provided. Furthermore, careful observation of the chromosomes in the wild-type and mutated flies suggested that each chromosome carries a specific set of genes, lying in a precise order along it. The banding pattern of the chromosome made it possible to begin the task of *mapping* – of identifying the exact position that a particular Mendelian gene occupied. A new research field, of *cytogenetics* – the cellular and microscopic study of genes – had been created.

SIMILARITIES AND DIFFERENCES

The term 'gene' now had two different meanings. On the one hand it was still an abstract entity, the determinant of a particular phenotypic character; on the other, it had a clear location, a 'map reference', and could be shown to be physically transmitted from cells to their offspring during both division and sex. And, as the international centre of gravity of genetics research had shifted to Morgan's lab, with its expertise in both development and genetics, there should have been the prospect of a synthesis between the two. It was not to be, and the reasons for the failure are significant. In part, they lay in the overriding differences in the biological questions that the two disciplines were asking. But also, genetics now seemed simple by comparison with developmental biology. Indeed, there was a strongly held view that one could and should bypass the sheer messiness of living organisms entirely, and their complex biochemistry, and focus instead on the mathematical niceties of breeding ratios, with their clear-cut experimental results. Organisms became merely probes with which to investigate genes.

The major concern of developmental biology remained the apparently inexorable programme that led from a single fertilized egg to the fully formed organism, the amazing sequences of cell division and migration, the partitioning out of an originally homogenous cell mass into defined structures, tissues, limbs. How is it that what seem at first sight to be very similar cell masses, going through seemingly similar transformations, end up in the one case producing a mouse and in the other a human? The similarity is so great that it led the Darwinian embryologist Ernst Haeckel to claim that 'ontogeny recapitulates phylogeny', in the belief that a human foetus in its pre-natal trajectory from fertilized egg to fully competent infant also traverses all the evolutionary steps that had led via fish, amphibian and reptile-like precursors to *Homo sapiens*.[13]

Why do the daughters of a cell from one part of the dividing embryonic cell mass end up as liver, and from another as brain or bones, with their very different pattern of proteins and characteristic shapes? How is it that all individual humans end up so astonishingly similar, so that nearly all of us as adults are between 1 and 2 metres tall, with two arms and two legs, and hands and feet carrying precisely five digits at their ends? Why do we have one heart but two lungs, and a single brain divided into two almost identical hemispheres? Such questions make developmental biology the science of the rules that produce regularities, *similarities* between organisms. For developmental biology, genes are seen not so much as isolated units but as part of a harmonious dialectic of interaction with the environment by which fertilized cells become mature adults through a trajectory described as *ontogeny*. And, as will become clear in due course, the constraints on this trajectory are only in part genetic.

By contrast, genetics was and is concerned not with similarities but with *differences*. Why is one *Drosophila* red-eyed, the other white-eyed? Why do people differ in height, and why do some have in their blood cells a haemoglobin molecule which seems unable to bind and carry oxygen as efficiently as it does in others? These *why* questions are to be answered in terms, ultimately, of the modern descendants of Mendelian determinants, the genes. Thus for genetics, genes are discrete units which lead in linear fashion, almost independently of one another and the environment in which they are expressed, to red or white eyes, for example, or to normal or sickle-cell haemoglobin. Ontogeny is of interest only in so far as genetic differences may produce abnormalities in development, such as the human pedigrees in which children are born with six-fingered rather than five-fingered hands. Otherwise, the geneticists' organisms are empty of time and internal content; there are only genes and phenotypes. They have no trajectory, no lifeline. If you want to see a good example of such empty organisms, turn to Chapter 3 of Richard Dawkins' *The Blind Watchmaker*, where he presents his computer game of how evolution may have worked, based on model organisms he calls 'biomorphs'.[14] Each biomorph springs fully formed from its predecessor. It has no development, no need to be subjected to the real constraints of growth, of ontogeny. Of course, we are not (I trust) intended to take such models too seriously, and my contrast between the interests of geneticists and developmental biologists is put in this stark form, which masks more subtle contrasts, because I contend that the difference in thinking between the two disciplines is real and has helped shape scientific history.

WHY GENES AREN'T 'FOR' ANYTHING

The step that took genetics beyond Morgan's location of genes to chromosomes also brought it into conjunction with biochemistry for the first time.[15] The organisms of choice were no longer *Drosophila* but even simpler organisms, initially the mould *Neurospora crassa*, responsible for the crust which forms on the surface of stale, damp bread, and later the common gut bug *Escherichia coli*. Mutations in these organisms were even easier to induce and study than in fruit flies, but now the consequences were no longer to be sought in phenotypic characters such as red or white eyes. Instead, they were metabolic. These organisms can be grown in covered saucers (called Petri dishes) filled with inert gels to which necessary food materials – sugars, amino acids or whatever – have been added. Whereas wild-type organisms can exist on very simple mixtures, some mutants cannot, but have to be fed with additional amino acids or other substances (*metabolites*). The wild-types can apparently synthesize such molecules from the chemicals provided to them, but the mutants can't. It turned out that what such mutants lack is specific enzymes which play a crucial role in the pathways that lead to the missing metabolites. Each specific mutation leads to the absence of a specific enzyme. So now a new and further definition of a gene became possible, and was formulated at Stanford in the 1930s by George Beadle and Edward Tatum on the basis of their experiments with *Neurospora*. One gene, they argued, equals, or produces, one enzyme, and they won a Nobel prize for the experiments that led to the formulation of this equation. And, in an odd way, for many researchers biochemistry now ceased to be a subject in its own right, but instead became merely a further tool – a technology – with which to study genes.

Genes themselves had taken a further step away from being hidden entities, unmoved movers. They could now not only be mapped on chromosomes, but also ascribed specific biochemical functions. They no longer determined characters, but instead, in a yet-to-be-understood manner, were responsible for the production of enzymes – perhaps even were enzymes themselves. This put a wholly new complexion on the meaning of a gene 'for' a character. Consider eye colour, for instance. The colour of the human iris depends on the presence in the cells of particular pigments. In the absence of pigment the eye is blue, and increasing quantities of the pigments provide shades which range from green to brown. Let us take for granted those developmental processes that lead to the formation of the eye, and within the eye the iris, and consider only the pigments themselves. Even if we ignore the biochemical steps whereby the necessary precursors to the synthetic pathway are produced, the direct pathway that leads to the synthesis of the eye pigments involves many different enzymes. Hence, on the Beadle–Tatum one gene, one enzyme principle, many genes must also be required (in fact, as we shall see, it is a great deal more complicated even than this).

So to biochemists, if not geneticists, there is no longer any gene 'for' eye colour. Instead there is a difference in the biochemical pathways that lead to brown and to blue eyes, for in the latter one particular enzyme, which catalyses a chemical transformation *en route* to the synthesis of the pigment, is lacking. So in blue-eyed people, the gene for this particular enzyme is either missing or non-functional for some reason. A gene 'for' blue eyes now has to be reinterpreted as meaning 'one or more genes *in whose absence* the metabolic pathway that leads to pigmented eyes terminates at the blue-eye stage'. Similarly, the reason for the difference in colour between Mendel's yellow and green peas is that the yellow ones have an extra enzyme in the metabolic pathway that leads to the breakdown of the green pigment chlorophyll. But this is of course just one of the many enzymes involved *en route* from the complex chlorophyll molecule to its end-products, to say nothing of the sequence of enzyme-catalysed reactions by which it is synthesized in the first place. This rephrasing yet again exposes the distinction between a developmental and a genetic approach. For the developmental biologist, what is of interest is the orchestrated biochemical route that leads to pigmented eyes. The mutation or absence of particular genes may help reveal that route (and is a technology in the sense that I define it in Chapter 3), but it is not of interest in itself; we are not dealing with one gene, one eye. But the geneticist is still interested in the difference between brown and blue eyes, yellow and green peas, and is still prepared to use the – misleading to the rest of the world and sometimes to geneticists themselves – shorthand of genes 'for' such colour differences.

Of course, all biologists know that this is true, and that the phrase 'genes for' is merely a convenient shorthand. Dawkins, in *The Extended Phenotype*, explicitly makes the same point, before going on to discount it as irrelevant provided the system behaves *as if* such 'genes for' existed. That is, his genes are purely theoretical constructs, combinations of properties which may or may not be embedded in specific enzymes or lengths of DNA, but which can be used to play mathematical modelling games.

You may think this doesn't matter, that to complain is merely pernickety pedantry on my part, but I assure you it is not. It matters a great deal. Thinking of genes as individual units which determine eye colour may not matter too much, but how about when they become 'gay genes' or 'schizophrenia genes' or 'aggression genes'? Sloppy terminology abets sloppy thinking. And it has implications for gene technology, too. As more is learned about the human genome, so early simplicities, such as the existence of a single gene responsible 'for' a particular disease, retreat. Many ostensibly 'single-gene disorders' are now known to result from different gene mutations in different people. All may show a similar clinical picture – for instance an inability to utilize cholesterol adequately, and hence have high levels of this substance in the circulating blood with consequently an enhanced risk of coronary heart disease. But the gene mutation, and hence the enzyme malfunction, that results in the disorder may be very different in each case. This also means that a drug which is effective in ameliorating the condition in one person may be simply ineffective in another in whom the cholesterol accumulation is the consequence of a different biochemistry. The implications for the utility of DNA testing are spelled out by Ruth Hubbard and Richard Lewontin:[16]

> . . . the patterns of transmission are unpredictable and seem to depend on various other factors, be they social, economic, psychological or biologic. The notion that health or illness can be predicted on the basis of DNA patterns becomes highly questionable. For each condition, extensive, population-based research would be needed in order to establish the existence and extent of correlations between specific DNA patterns and overt manifestations over time. Furthermore the correlations are likely to have only a degree of statistical validity, not absolute validity.

GENES BECOME DNA

We now come to the part of the story whose unravelling is regarded, with justice, as one of the great scientific triumphs of the century: the identification of the genetic material itself and the elucidation of what exactly is meant, in biochemical terms, by the Beadle–Tatum formulation of one gene, one enzyme.

Genes, as physical entities, lie on chromosomes in the nucleus. So it made sense to see what chromosomes are made of. This was not difficult. They could be shown to be largely composed of a particular class of protein (called histone), bound tightly to a seemingly inert long chain molecule, of a type which had originally been isolated in 1868 by the chemist Friedrich Miescher in Tübingen from pus collected from discarded surgical bandages. Miescher called the material *nuclein*, and later showed it to be highly enriched in other, less unsavoury sources of cell nuclei, such as salmon sperm. Nuclein was shown to exist in two forms, as deoxyribonucleic acid (DNA) and ribonucleic acid (RNA); and, after an initial confusion in which it was believed that the one was present in animal cells and the other in plant cells, it was realized that both were universally present in all cells. The DNA is almost entirely confined to the nucleus (though some is present in mitochondria); RNA is in both the nucleus and the cytoplasm that surrounds it.

Much was already known about proteins, but rather little about nucleic acids, and the consensus view during the 1930s and 1940s was that the active constituents of the chromosomes would turn out to be the proteins. A number of experiments seem with hindsight to have pointed conclusively in the opposite direction, but such was the power of the protein paradigm[17] that they were largely ignored or misinterpreted.

When the breakthrough came, it was from neither biochemical nor genetic research, but from an entirely unexpected quarter. In the early 1950s James Watson, an ambitious and bumptious young American post-doc on a visiting fellowship to Cambridge, and a brilliant but somewhat dilettante ex-wartime engineer and physicist, Francis

Crick, were attempting, with limited success, to identify the three-dimensional structure of DNA by means of the then relatively new and arduous techniques based on crystallographic analysis. Illumination arrived in the form of X-ray diffraction pictures, taken by Rosalind Franklin in London (Figure 3.4 on p. 62) and provided to the Cambridge pair without Franklin's knowledge. The pictures were the technology Watson and Crick required, for they immediately provided the clue to the now famous double-helix structure of DNA, and to the fact that its component *nucleotides* (sub-units) – adenine, guanine, cytosine and thymine – could fit together only within particular configurations which pointed unmistakably to how chromosome duplication and copying could occur. The structure is now universally familiar, but worth showing once more here (Figure 5.4).

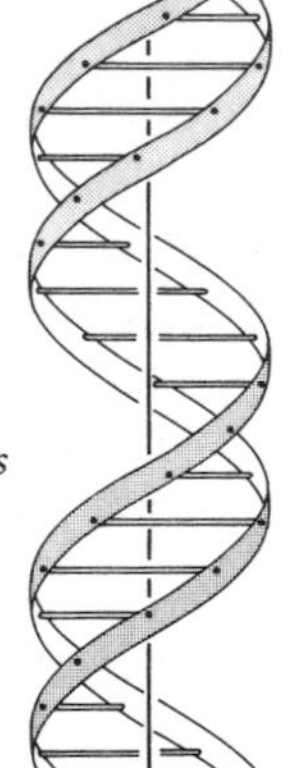

Figure 5.4 Watson and Crick's drawing of the DNA double helix.

As Watson and Crick saw, and as implied in their famous *Nature* paper, if the two strands of DNA were to unwind, each could provide the template on which its matching strand could be copied, without error. Hence identical sets of DNA strands – chromosomes – could be synthesized during mitosis and distributed to the daughter cells. They concluded thus:[18]

> It has not escaped our notice that the specific pairing we have postulated immediately suggests a possible copying mechanism for the genetic material.

So the 'gene' had once more been transmuted. It could now be considered as being constructed of DNA. But what made a length of DNA a gene? By this time Beadle and Tatum's formulation of one gene, one enzyme had been slightly broadened. Enzymes are proteins (mainly), but not all proteins are enzymes. Some, like the microtubular protein tubulin, form the structural skeleton of cells; others, like collagen in connective tissue, fill the spaces between cells; still others, like haemoglobin in blood, fulfil vital but non-enzymic metabolic functions. So it would be better to speak of one gene, one protein. Or, even more precisely, as proteins can be built of several amino-acid chains (polypeptides) cross-linked or otherwise bound together, one gene, one polypeptide chain.

A brilliant decade of theory and experiment – the biologist Gunther Stent has called it the classical age of molecular genetics[19] – seemed to provide many of the answers. By the mid-1960s a startlingly simple picture had emerged, and genetics and biochemistry had combined into the new science of molecular biology. DNA is composed of four nucleotide bases (abbreviated to A, C, G and T), there are 20 erent naturally occurring amino acids. The physicist George Gamow treated the problem as one of code-breaking. If it took two bases to code for each amino acid, there could only be 16 (4 × 4) possible combinations, which was too few. If three, there were 4 × 4 × 4 or 64 possible combinations, which was too many, but would serve if the code were redundant (more than one triplet combination for any one amino acid) and also if some triplet combinations of bases had other 'meanings' for the code, such as signalling 'start here' or 'stop here'. So it proved, as elegant experiments succeeded in reading the DNA code, and matching triplet sequences to specific amino acids.

And a further, vital complexity. There are, as mentioned above, two forms of nucleic acid in cells. One is DNA, the other the closely related but single- rather than double-stranded RNA. In eukaryotes DNA is present in the cell nucleus, where it comprises the chromosomes. RNA is present in both the nucleus and the cytoplasm. It transpires that proteins are actually synthesized in the cytoplasm, and the copying procedure for DNA during protein synthesis consists of first the partial unwinding of the DNA double helix, then the copying of a single strand of RNA (called messenger RNA). The RNA then moves out from the nucleus into the cytoplasm, where it provides the template for the synthesis of particular protein chains.

So, a gene was now a length of DNA, a region of a chromosome which can be copied into RNA which in turn codes for – that is, provides a template for – the stringing together of the sequence of amino acids that makes a protein. Furthermore, this synthesis is a one-way street. A sequence of amino acids in a protein cannot serve as a template for the synthesis of RNA and thence DNA. Crick, whose sense of the *mot juste* has remained with him in the more than forty years since the double helix was first presented to the world, has called this the 'Central Dogma' of molecular biology, a one-way flow of information:[20]

$$\text{DNA} \rightarrow \text{RNA} \rightarrow \text{protein}$$

. . . once 'information' has passed into the protein it *cannot get out again.*

This formulation, as will become clear in Chapter 7, is as central to ultra-Darwinian theory as it is to molecular biology. And to continue the linguistic, information-theory metaphor within which genetic theory was now to be formulated, the directed synthesis of RNA on DNA was termed *transcription*, and the synthesis of protein on the RNA was *translation.* DNA had become the master-molecule, and the nucleus in which it was located had assumed its patriarchal role in relationship to the rest of the cell.[21] It is hard to know which had more impact on the future directions of biology – the determination of the role of DNA in protein synthesis, or the organizing power of the metaphor within which it was framed.

The fact that the development of computer technology, with its demands on information theory, has occurred contemporaneously with the growth of molecular biology has not merely provided the physical technology, in instrumentation and computing power, without which the dramatic advances of the decades since the 1960s would not have been possible. It has also given the organizing metaphors within which the data are analysed and the theories created. Crick may have originated the metaphor, but it has taken Dawkins to draw it to its logical conclusion. Consider for example, the euphoria of this account in *The Blind Watchmaker* in which he considers a willow tree in seed outside his window:[22]

It is raining DNA . . . It is raining instructions out there; it's raining tree-growing, fluff-spreading algorithms. That is not a metaphor, it is the plain truth. It couldn't be any plainer if it were raining floppy discs.

This is fine writing, great fun to read, so much so that it has found its way into anthologies of scientific prose. But it is misleading in almost every respect. You might ignore the trivial fact, irritating to a biochemist like myself but airily dismissed in the paragraph containing this extract by the grand theorist, that the seeds contain a great deal more than DNA: there are proteins and polysaccharides and a multitude of other small molecules without which the DNA would be inert. But you cannot ignore the blunt statement that 'this is not a metaphor', for this is precisely and at best what it is. It certainly isn't 'the plain truth'. Nor is it a statement of homology or analogy. It is a manifesto.

In his more recent book *River out of Eden*, Dawkins is more explicit still. Living organisms may be regarded as analogue devices, he argues, but the analogue machines that are us are constructed and directed by DNA, which is essentially digital. The information content of the genome can be expressed in terms of bits and bytes. And what is life but an expression of the working-out of the genome (or, as we shall see in Chapter 7, the genome's way of replicating itself)?

Life is just bytes and bytes and bytes of digital information.[23]

Mendel has been solved, and turned into chemistry plus information theory. Right? *Wrong*: Dawkins may regard himself as nothing but a digital PC, and his complex lifeline in space and time as the read-out from a one-dimensional string of A's and C's and G's and T's, but things are a bit more complicated than this, both for him and for any other living organism.

RUSSIAN DOLLS

Periods of great unifying simplicity in science are frequently followed by times in which simplicity dissolves once more into complexity. So it has been for molecular biology since the 1960s. The problems begin with a simple conundrum: the amount of DNA in the chromosomes of any organism turns out to be far, far greater than can be accounted for by a simple calculation based on numbers of proteins and a triplet code. As the average protein is perhaps 300 amino acids long, it requires a length of DNA of 900 nucleotide bases to code for it. And as humans are estimated to express perhaps 100,000 different proteins in the varying tissues of the body and throughout their lifeline, then the human genome should consist of some 90 million DNA bases (or rather, base pairs, to describe the fact that each base has its match in the other chain of the DNA double helix), distributed, like strings of beads in a necklace, among the 23 human chromosomes. But in fact the armies of molecular biologists engaged in the vast task of sequencing the human genome are faced not with a mere 90 million, but with 3 billion base pairs – a more than thirty-fold excess. What is all this extra DNA about?

Part of the answer was known pretty early on. It's no good just having a set of genes for proteins and expecting that this is all that is required to build an organism. As we have seen, not all proteins are being synthesized at all times in all cells. Indeed, during development, as cells in the developing embryo lose totipotency and become specialized, some genes must be, so to say, switched off, and others switched on, depending on the fate of the particular cell. Nerve cells, but not liver cells, need to be able to synthesize neurotransmitter molecules. There must therefore also be a set of instructions to the genes to switch them on or off at appropriate times. If, as digital theory and the Central Dogma insist, these instructions must ultimately come from the DNA itself, then there must also be another class of genes present, not coding for proteins but acting as on or off switches. Such switch genes (*operons* and *repressors*) were first identified in bacteria in the 1960s by Jacques Monod and François Jacob working in Paris. Variants of them function in all organisms, prokaryotes and eukaryotes, single and multicellular organisms, so some of the extra DNA is accounted for by these regulatory functions (the molecular mechanisms by which they work need not concern us here).

But even when these additional functions are accounted for, well over 90 per cent of the DNA of the human genome has no known function. Much of this DNA consists of repeating sequences of bases, and hence has been called repetitive DNA; more arrogantly, molecular biologists refer to it disparagingly as 'junk', or, for reasons which will become apparent in Chapter 8, 'selfish' – a term due I believe to Crick, who himself evoked it as a deliberate echo of Dawkins' 'selfish gene'. (Note that Crick's DNA's selfishness is demonstrated by the fact that it doesn't do anything for the cell or the organism in which it is embedded; it simply allows itself to be copied. Dawkins' selfish genes, on the other hand, are selfish because they specifically aid the successful reproduction of the organism that contains them, and hence promote their own replication.) It is this selfish DNA that the international teams of highly skilled sequencers employed within the Human Genome Project are painstakingly working through at a cost originally estimated as a dollar a base – a task that Watson once notoriously dismissed as fit only for trained monkeys.

If this were the only problem, simple-minded 'gene = DNA' theory might not be in too much trouble. But there is more, much more. First, as the mapping and sequencing tasks have proceeded it has become very clear that the DNA beads-on-a-chromosome-string view of genes is too simple. Individual proteins turn out not to be coded for by a simple continuous strand of triplet bases. Instead, the coding regions of the DNA are interspersed with vast tracts of repetitive, non-coding regions which have been given the name *introns*. Nor are the coding regions arranged sequentially, so that all that would be necessary to read them would be to 'airbrush out' the intervening regions. Instead, different parts of a protein may be coded for by segments of DNA distributed across long regions of the chromosome, and brought together by complex cellular machinery – a process known as *splicing*.

But once splicing is possible, it also becomes possible to arrange the spliced sequences in a variety of ways which are not automatically to be read off from the originating DNA. Many proteins are the products of such alternative splicing arrangements. Many more are synthesized on DNA in one form and subsequently processed further in the cell, having components added or removed – a process known, in a continuation of the computer and linguistic metaphors, as *editing*. And, as you may be expecting, there are alternative editing processes. The result is that, far from being able to speak of one gene, one protein, both genes and proteins are disarticulated. Genes can be assembled from alternative pieces of DNA or rearranged so that their codes are read differently (Figure 5.5). And proteins take on multiple forms as a result of cellular processes a long way downstream from DNA itself. The term 'gene' in the original Mendelian sense, or in the Beadle–Tatum sense, no longer means quite the same as 'DNA strand on a chromosome'.

And even now we aren't quite finished. Aeons before the new molecular biology arrived on the scene, back in the days of cytogenetics in the 1930s, Barbara McClintock was studying the genetics of maize – not as theoretically fashionable as Morgan's *Drosophila*, but of

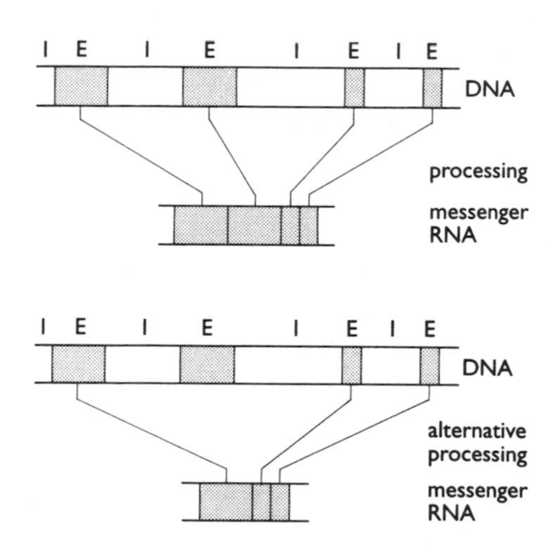

Figure 5.5 Introns (I), exons (E) and alternative splicing of messenger RNA.

a good deal more immediate practical significance to the agriculture of the US Midwest's corn belt. Unlike Mendel, McClintock was no maverick outsider – she was only the third woman ever to be elected to the US National Academy of Science (in 1944) and went on to become President of the Genetics Society of America in 1945, but what she identified as occurring in her maize chromosomes was so far outside the conventional wisdom of the time that it was ignored or suppressed for as long as Mendel's ratios. In the end, McClintock was more fortunate than Mendel, for she lived long enough to see her findings rehabilitated within the new framework of complexity in the 1980s, and she herself, following a widely acclaimed biography,[24] was finally awarded a Nobel prize in 1983.

McClintock's original observations, and the reasons why they were rejected for so long by orthodox genetics, have been hotly debated; be that as it may, what she actually observed is now accepted as having transformed the static concept of the genome that had hitherto dominated genetic thinking. I have already described the characteristic appearance of chromosomes, as twisted ribbon-like structures with irregular but precise patterns of bands running across them. Although during sexual reproduction matching sections of each pair of chromosomes can exchange, thus enhancing the genetic shuffling that occurs with sex, the patterns that result, and which indicate the location of any particular gene along the chromosome, were thought to be stable. McClintock's finding destabilized the genes. It appeared that, although stability was the general case, genes could also 'jump', relocating themselves at different sites in the chromosome map. Amply confirmed by the molecular biologists in the post-Central Dogma climate which was beginning to emerge in the 1980s, jumping genes became no longer at best a special case in maize, but part of what increasingly has to be perceived as a fluid rather than a stable genome. Far from being isolated in the cell nucleus, magisterially issuing orders by which the rest of the cell is commanded, genes, of which the phenotypic expression lies in lengths of DNA distributed along chromosomes, are in constant dynamic exchange with their cellular environment. The gene as a unit determinant of a character remains a convenient Mendelian abstraction, suitable for armchair theorists and computer modellers with digital mind-sets. The gene as an active participant in the cellular orchestra in any individual's lifeline is a very different proposition. I have summarized the differences (some of which will become clearer in the chapters that follow) between the abstract genes of the theoreticians with the real-life genes of the biochemist and molecular biologist in Table 5.1.

Table 5.1 Theoreticians' genes versus biologists' genes

Theoreticians	*Biologists*
Gene as a theoretical entity	Gene as a term applied to varying sequences of DNA
Genes seen as unitary and indivisible, rather as atoms were before the days of nuclear physics	Genome fluid: DNA strands subject to alternative reading frames, splicing and editing processes
Beanbag models of gene expression	Gene expression contingent on cellular regulation at levels from the genomic to the organismic
Assumption of linear one-to-one relationship between genotype and phenotype	Relationship between genotype and phenotype sometimes linear and one-to-one, but this is a special case of a more frequent norm of reaction
'Preformationist' assumption of 'empty organism' that ignores developmental trajectories	The ontogeny of information
Genetic primacy: deviations are 'phenocopies' or modelled by incomplete penetrance or partial dominance	Some phenotypic conditions mimicked by genetic conditions, e.g. schizophrenia, breast cancer, Alzheimer's ('genocopies')

GENES AND CELLS

So what part in the symphony does DNA play? The actual nucleic acid macromolecule is really rather boring. Step into San Francisco's hands-on science museum, a vast aircraft hangar of a place called the Exploratorium, and amid the cacophony of sound and flashing lights that fill the space you will find a small undistinguished display, a beaker half full of a clear, thick liquid with a glass rod dipped into it. The liquid is a concentrated solution of urea containing dissolved DNA. Pull the rod slowly out, and a whitish thread trails behind it into the surface of the solution. This is a pure DNA fibre, and you are mimicking the procedures by which Miescher originally purified it, for it is surprisingly stable and inert. Few proteins could withstand the chemical brutalizing required to isolate DNA. Of course, if DNA weren't so stable, the plot of *Jurassic Park* would be even more improbable than it already is. DNA molecules really can survive for long periods, much longer than proteins can, even though the claims that they can be extracted from fossil material or insects embedded in amber have been recently discounted.

What brings DNA to life, what gives it meaning, is the cellular environment in which it is embedded. Watson and Crick's great insight that, because of the structure of DNA, if the two strands of the double helix were unwound then each could provide the template on which a second matching strand could be built, gives the impression that such a process is a simple bit of chemistry. Genetic theorists with little biochemical understanding have been profoundly misled by the metaphors that Crick provided in describing DNA (and RNA) as 'self-replicating' molecules or replicators, as if they could do it all by themselves. But they aren't, and they can't. Replication isn't an inevitable chemical mechanism. You may leave DNA or RNA for as long as you like in a test-tube and they will remain inert; they certainly won't make copies of themselves. To perform the copying, is it not sufficient for the cell to have available the necessary precursor molecules, the font of A's and C's and G's and T's, each requiring their own painstaking synthesis from even simpler substances. In addition, particular enzymes are required to unwind the two DNA strands, and others to insert the new nucleotides in place and zip them up again. And the whole process requires energy, the expenditure of some of the cell's ubiquitous ATP. While chemical synthesizers designed by human instrument-makers can now provide the technology to build defined artificial DNA sequences at the behest of the biotechnologist, the cellular processes involved are far from trivial.

So, of course, are the steps that result in the synthesis of particular proteins based on the DNA. The histones surrounding the relevant region or regions of the double helix must be unwrapped, the DNA strands must be separated, enzymes must transcribe the 'sense' strand into its matched length of RNA, and individual RNA lengths must be spliced, edited and further manipulated in the cell nucleus. Even then, there are further controls. To leave the nucleus and be inserted into the copying machinery in the cell cytoplasm, the RNA message must pass through the nuclear membrane, for which it requires a biochemical 'exit permit', provided by the membrane proteins. In eukaryotic cells, this ribosomal machinery itself consists of a giant assemblage of sub-units together containing more than 80 different proteins, and RNA sequences containing more than 6,700 nucleotide bases. Without it, without the complex biochemical environment the cell provides, 'genes' in the DNA sense of the term, simply can't function.

To appreciate the significance of this, consider viruses, which consist almost entirely of DNA (or in some cases RNA) surrounded by a protein coat, and capable of being stored indefinitely as elegant crystalline solids (Figure 5.6). What brings the virus to life is the property provided by its proteins of being able to penetrate the membrane of a victim cell and release its own DNA, which parasitizes the cellular replicative machinery of its hapless host. The host cell is thus forced to copy and translate the viral DNA as if it were its own, filling itself with newborn virus particles until it bursts. The newly hatched viruses are thus released into the environment where each, once brought into contact with a fresh prey cell, can repeat the sequence. Viruses are often described as the most basic of living forms, 'naked replicators'. The question of whether they are or are not 'alive' depends on how you define life, and may be a matter of mere semantics;

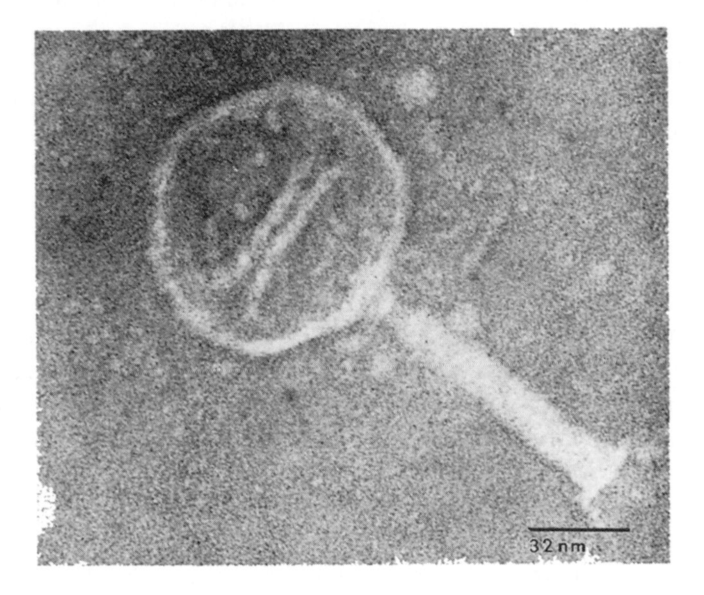

Figure 5.6 A single virus – a nucleic acid head surrounded by a protein jacket and tail.

that they cannot replicate except within a cell which *is* clearly alive[25] is certain. Naked and alone, they are powerless to act.

This is why an individual's lifeline requires more than the mere mixing of parental DNAs at the moment of fertilization. Sperm, to be sure, are somewhat like viruses, in that they provide only DNA. But an egg contains more than just the maternal complement of DNA to match that provided by the paternal sperm. It has in addition all the cellular apparatus required to bring both sets of the DNA together and persuade the otherwise inert fibres to play their part in the cellular orchestra. Among the major contributions from the maternal cytoplasm are mitochondria which generate the necessary energy supply for the orchestra to start up, and which in addition carry an independent set of DNA molecules, quite separate from those in the nucleus, whose evolutionary significance will become apparent in Chapter 8. This asymmetry between the two sex cells, egg and sperm, at the very moment of conception is of profound developmental and evolutionary significance.

From this moment of conception on, the maternal cellular machinery is responsible for directing the activation of particular genes (DNA sequences) and hence the synthesis of specific proteins. These proteins in turn include some whose function is to act as switches – regulators to turn on, and in due course turn off, other DNA sequences. A continuous cycle of synthetic activity begins in which DNA sequences are uncovered, transcribed into RNA, processed, spliced, edited and translated into proteins, which then provide feedback control to the DNA, perhaps switching off their own synthesis, perhaps switching on the synthesis of other proteins by uncovering other DNA sequences or influencing the splicing and editing steps. This exquisitely timed and subtly orchestrated cellular symphony culminates in due course in the synthesis of the proteins that begin the process of replicating and segregating the chromosomes once more, enabling the cell to divide and the cycle to recommence. This is why Dawkins' claim that his garden willow tree is simply 'raining DNA' is so biochemically wide of the mark.

In the digital information metaphor, these cellular mechanisms play no part in the creation of this symphony. They are as dumb as the mechanism by which a cassette player converts the trace on a magnetic tape into a Beethoven violin concerto or a Miles Davis jazz track. All that the tape head and the speakers do is to follow the instructions given by the tape. They can influence the quality and fidelity of the sound that is emitted, but they don't carry information. The symphony remains in the DNA. But this is not how cells work. Unlike the cassette player, they don't merely play their 'tape' at constant speed and hang the consequences. They instruct the tape as to which bits to play and when to play them, and they also edit the output. And of course, also quite unlike the cassette player, they continually reconstruct themselves throughout the cell cycle and the lifetime of the organism which they comprise. In so far as the information metaphor is valid at all, it can be expressed only in the dynamic interaction – the dialectic, therefore – between the DNA and the cellular system in which it is embedded. Cells make their own lifelines.

GENES, ENVIRONMENTS AND NORMS OF REACTION

Thus in both the Mendelian and the biochemical senses, genes are only partially determinate entities within genomes. They are not independent beads on a necklace. This is why it has become a modern convention to speak instead of the genome as fluid. How, when and to what extent any gene is expressed – that is, how its sequence is translated into a functioning protein – depends on signals from the cell in which it is embedded. As this cell is itself at any one time in receipt of and responding to signals, not just from a single gene but from many others which are simultaneously switched on or off, the expression of any single gene is influenced by what is happening in the whole of the rest of the genome.

So when we talk glibly about the development of an organism being 'a product of the interaction of genes and environment', the phrase masks as much as it reveals. Neither gene nor environment, as we have seen, is an unproblematic term. First, a 'gene' as an abstract determinant is quite different from the complex processing mechanisms that put together the particular DNA sequences that define the primary sequences of proteins. Nor, of course, are proteins merely defined as their primary sequences. As already discussed in the context of natural kinds in Chapter 2, they have complex secondary and tertiary structures which depend not just on their amino-acid sequence but on their environment, on the presence of water, ions and sometimes other small molecules, and on acidity or alkalinity. The path from primary structure to fully fledged protein does not contain as many regulatory steps as that from DNA to protein, but it does involve orders of complexity which move us yet further from the one gene, one protein heuristic. And as proteins themselves become assembled into higher-order structures within the cell, still more constraints come into play.

The school textbooks which start with Mendel and his ratios have it wrong. Without Mendel, genetics would never have got off to such a flying and seemingly straightforward start, and he deserves to be honoured for his experiments. But the founders of a field, by choosing experimental systems which seem to give clear-cut answers, often also produce an appearance of simplicity which is ultimately misleading. The famous and paradigmatic Mendelian ratios are the results of rather special cases, the phenotypic expressions of enzyme pathways rather little influenced by environmental circumstance, perhaps just because they reflect relatively trivial features of that phenotype. By contrast, the expression of most genes is modified at several levels. It is affected by which other genes are present in the genome of the particular organism, by the cellular environment, by the extracellular environment and, in the case of multicellular organisms, by the environment outside the organism.

An example. Gene technology is now so advanced that it is possible to generate virtually at will ('construct' is the somewhat odd term the geneticists use) organisms – mice for instance – into which particular genes have been inserted, or from which they have been deleted. Of course, many such constructed mutations are lethal, and the embryos that carry them are either spontaneously aborted or can survive for only a few days or weeks. These monstrous births – such as the so-called oncomouse, which carries a mutation that results in the animal developing a cancer – have for obvious reasons been the source of much legal and ethical heart-searching. But the point I wish to make here is a different one. In quite a number of cases where genes coding for proteins which are supposed to have vital functions within the cellular economy have been deleted (so-called 'knock-out mutants'), the absence both of the gene and of the protein whose synthesis it codes for seem to make little observable difference to the life of the animal. It has, as they say, an apparently normal phenotype.

Does this mean that the original view, that the protein concerned played a vital role in the cellular economy, was false? Not at all. It is a demonstration instead of the power of developmental *plasticity*, the capacity of a living system to adapt to experience and environmental contingencies, and to compensate for deficiencies. This capacity is augmented by the functional redundancy present in all organisms. Redundancy assists stability; it means that there may be many alternative routes that the cell and the organism can adopt during development which can lead to an essentially identical end-point. In the presence of a particular gene and protein, one route is adopted, and in their absence another is taken. Once again, there is no necessary linear path between gene and organism. It is interesting that such redundancy is now recognized by engineers as a feature of good design in human technology too.

But such plasticity is of course not infinite: there are sharp limits to the tolerance of any gene – or any phenotype – to environmental change. Outside these limits, the response is to curl up and die. But within them, the expression of any gene may be defined in terms of its *norm of reaction* to the environment – a term originally introduced by the population geneticist Theodosius Dobzhansky in the 1950s, and rather out of fashion with today's theorists. In the beanbag thinking that follows the Mendelian tradition, for any one gene there is only one phenotypic outcome. By contrast, in Dobzhansky's concept of norm of reaction, the phenotypic expression of any gene may vary over a wide range, depending on the environment in which it is being expressed. And remember, that environment includes the products of all the other genes in the organism's genome, as well as external factors impinging upon it.

Recognizing that there is no linear relationship between gene and phenotype, E.O. Wilson, the founder of sociobiology, speaks of 'genetic tendencies', 'predispositions' or 'inclinations', and prefers as a metaphor the thought that 'genes hold culture on a leash'.[26] The metaphor simultaneously privileges the gene as once more an unmoved mover, while bowing to the inevitable of non-linearity. It is far more appropriate to recognize, as Dobzhansky did, that genes and environments are dialectically interdependent throughout any individual's lifeline, that the argument for primacy is a reversion to an almost pre-scientific doctrine of preformationism which we can surely now transcend. Our science should be adult enough to rejoice in complexity.

NOTES

1. Some people would say too lucky; his published ratios have been subjected to statistical re-evaluation, and on this basis seem too good to be true! It is as if he knew what he wanted to find, and encouraged the data to reveal it. Repetitions of his experiment are unlikely to yield such unambiguous results.
2. Peter J. Bowler, *The Mendelian Revolution*.
3. Respectively Carl Correns, Erich von Tschermak and Hugo de Vries.
4. Richard Dawkins, *The Extended Phenotype*.
5. William B. Provine, *The Origins of Theoretical Population Genetics*.
6. Donald MacKenzie, *Statistics in Britain, 1865–1930*.
7. See e.g. K. M. Ludmerer, *Genetics and American Society*.
8. The sequence is well described by Lewis Wolpert in *The Triumph of the Embryo*, though his claim that the whole process is formally reducible to an unrolling of a DNA-based programme is not an interpretation I share, as the next chapter will make clear.
9. Donald Fleming, in his introduction to Loeb's *The Mechanistic Conception of Life*.
10. Dawkins, *The Blind Watchmaker*, p. 37.
11. Lily E. Kay, *The Molecular Vision of Life: Caltech, the Rockefeller Foundation and the Rise of the New Biology*.
12. Garland E. Allen, *Thomas Hunt Morgan*.
13. See Stephen Jay Gould, *Ontogeny and Phylogeny*, for the history of this claim.
14. Dawkins' computer games have now (1996) been spun off as a separate product, *The Evolution of Life*, available as a CD-ROM.
15. This isn't quite true: Garrod's work on inborn errors linked genetics and biochemistry in the 1900s, but, perhaps because it was closer to clinical than to basic research, it tended to be ignored.
16. Ruth Hubbard and R.C. Lewontin, 'Pitfalls of genetic testing'.
17. Robert Olby, *The Path to the Double Helix*.
18. James D. Watson and Francis H. C. Crick, 'Genetical implications of the structure of deoxyribonucleic acid'.
19. Gunther Stent, *The Paradoxes of Progress*, Freeman.
20. Olby, *The Path to the Double Helix*, p. 432.
21. Bonnie Spanier, *Im/partial Science*.
22. Dawkins, *The Blind Watchmaker*, p. 111.
23. Dawkins, *River out of Eden*, p. 19.

24. Evelyn Fox Keller, *A Feeling for the Organism: The Life and Work of Barbara McClintock.*

25. Bill Pirie points out that certain viruses may be able to replicate in killed cells.

26. Edward O. Wilson, *On Human Nature*, p. 172.

6

Lifelines

Life is the expression of a particular dynamic equilibrium which obtains in a polyphasic system.

Frederick Gowland Hopkins,
'The dynamic side of biochemistry'

ORGANISMS IN FOUR DIMENSIONS

At the heart of modern biology lies the issue of the nature of individual living units – organisms. Notwithstanding the cautionary words of Chapter 2 on the ambiguities latent in our sense of the borderlines between ourselves and the external world, for most of the time we all have a sense of our own existence as a coherent whole, and we recognize such coherence and unity in others – and not just of our own species. Dog and frog, worm and amoeba – each has a recognizable existence as an individual organism. So does an oak tree and a marigold, though our picture may get a little hazy when it comes to considering the spreading clumps of buttercups in the lawn or the mushrooms beneath the tree.

Organisms differ dramatically in scale, from blue whales to bacteria, but every one, large or small, exists as a three-dimensional object occupying a defined volume within its environment, and each possesses recognizable structures, internal features and organization. But these three dimensions of space cannot provide a full description of an organism, for it extends in time as well as in space. It may begin by budding off from a pre-existing single-celled organism, like a yeast cell. Or it may grow and develop: oaks from acorns; humans, dogs and frogs from the fertile combination of egg and sperm. Some organisms seem essentially fully developed – mature – at the instant of their formation, as with the newly budded yeast. Some develop incrementally over time, growing throughout their life, as do many trees. Others, such as ourselves, grow for a period, and reach a seemingly stable mature stage before ageing and beginning to decay. Yet others go through a series of radical transformations in which entire body-plans are reconstructed, as when eggs become caterpillars become chrysalises become butterflies. As with space, so with time, living forms range through many orders of magnitude, from the bacteria whose time between divisions may be only twenty minutes to the thousand-year-old giant redwoods of California.

The time dimension can never be ignored. Life persists not in three but in four dimensions – persistence which depends above all on the maintenance of order: order within the cell, order within the organism, order in the relationship of the organism to the world outside it. It is the meaning and mechanism of this persistence, the generation and maintenance of both short- and long-range order, which form the theme of the present chapter. Genes and genomes neither contain the future of the organism, in some preformative modern version of the homunculi van Leeuwenhoek thought he saw in the sperm, nor are they to be regarded, as in modern metaphors, as architects' blueprints or information theorists' code-bearers. They are no more and no less than an essential part of the toolkit with and by which organisms construct their own futures.[1]

CELLS, ORGANISMS, ENVIRONMENTS

Neither cells nor organisms can be considered in isolation from their own external environments. All cells are surrounded by membranes, constructed of complex arrays of lipid and protein molecules, which act as both barrier and interface with the world outside them. Across this semipermeable barrier there is a constant traffic with the cell's surroundings. To survive, let alone to act upon the external world or to replicate, requires the continual expenditure of energy, energy derived from food in the form of pre-existing molecules such as sugars or fats, or, for green plants, by photosynthetic processes which rely on carbon dioxide and water. All these molecules must be carried into the cell across its membrane, and waste metabolites ejected through it into the environment. But the membrane has to be selective: while letting in desirable substances, it has to do all it can to keep out those which could be harmful.

For single-celled organisms, the environment of the cell is obviously also that of the organism, the ever-fluctuating external world, inherently patchy. Some parts of that world may be antithetical to survival – too hot, too dry, too acid. Some may be rich in food sources, others poor. Supplies of potential food may vary: in one area glucose may be abundant, in another a different sugar. Faced with such patchiness, many single-celled organisms can take steps to seek out more favourable conditions, especially if they are in a watery environment. Not all are content to go with the flow. The membranes of many species are equipped with chemosensors enabling them to detect gradients of concentration of sugar solutions, and tails (flagella) or oars (cilia) enabling them to swim or row up the gradient to regions richer in foodstuff. Similarly, they may move away from regions which are too acid or too hot.

But their power to choose a favourable environment is limited by the range of environments accessible, and survival will also depend on the ability of the organism to adapt to less than optimal conditions. If one food supply is absent but another potential source is available, single-celled organisms may need to produce the enzymes necessary to digest what is available. This is indeed what Monod and Jacob found in their experiments which identified the operon: bacteria which did not normally possess the enzymes required to metabolize the sugar lactose would synthesize them if their food supply was restricted only to lactose. This doesn't require an organism to invent from scratch the DNA that could be used to direct the synthesis of a novel enzyme – that would be beyond the capability of a single cell within its lifetime. The bacteria already contain the DNA sequences necessary to produce a lactase enzyme, but in normal circumstances these are switched off, and are turned on only by signals triggered within the cell by its sensing the lactose-rich, glucose-poor environment across its membrane. It is the organism in interaction with its environment, therefore, that determines which of its available genes are to be active at any one time.

For a multicellular organism, such interactions between cells and environments are more complex. Individual cells no longer have to work in isolation, exposed to the Great Outdoors. Rather, each is surrounded by its own microenvironment, external to the cell but internal to the organism. It is the organism as a whole that has to respond to the patchiness of its environment so as to optimize its life-chances. The cells within are buffered from the wilder external excesses, bathed in an extracellular fluid whose temperature and composition remain comfortably constant, as near-optimal as possible for the cells it surrounds, wafting them food and oxygen and washing away their unwanted excretions. Such cosseted creatures no longer have to be constantly on the look-out for uncertain food supplies, their genes in a state of readiness to make the switch from glucose to lactose, so they have no need to maintain a DNA repertoire that will enable them to make such a switch. The demands on them are simpler and more predictable.

There is a price to pay for the simple life. The individual cells lose their autonomy within the greater unity of the organism: they surrender their capacity for independent and unrestrained replication, and their totipotency. They become specialized, as liver or brain, leaf or root. In the course of this specialization, as ontogeny proceeds, particular DNA sequences are switched on or off in defined temporal sequences. It is no longer only a case of proceeding through the cell cycle to division, but of establishing cells with an appropriate structure, shape and pattern of enzymes to function as part of a particular organ. To ensure harmony at a multicellular rather than a cellular level, each cell has to be able to respond to the presence of its neighbours and to signals from distant parts of the organism arriving at its membrane surface with as much sensitivity as, within the cell, DNA sequences can respond to protein signals. The external membranes of individual cells within multicellular organisms are packed with specialized receptors which can respond to circulating signal molecules (such as hormones), and are punctuated with convoluted channels which permit the entry or exit of designated substances only. The cellular lifeline has become subordinated to that of the organism.

Like 'gene', the term 'environment' is thus complex and many-layered. For individual gene-sized sequences of DNA, the environment is constituted by the rest of the genome and the cellular machinery in which it is embedded; for the cell, it is the buffered milieu in which it floats; for the organism, it is the external physical, living and social worlds. Which features of the external world constitute 'the environment' differ from species to species; every organism thus has an environment tailored to its needs. As I shall argue in later chapters, organisms evolve to fit their environments, and environments evolve to fit the organisms that inhabit them. No environment is constant over time. Even for the individual gene, the genomic background against which it is expressed changes during the cell cycle as other genes are switched on and off. Outside the organism, change is virtually the only constancy. Stasis is death.

There are two lessons to be drawn from such descriptions. The first is that the boundaries between organism and environment are not fixed. Organisms are constantly absorbing parts of their environment into themselves as food, and are constantly modifying their surroundings by working on them, by excreting waste products, or by modifying the world to suit their needs, from birds' nests to beaver dams and termite mounds. Organisms – any organism, even the seemingly simplest – and the environment – all relevant aspects of it – interpenetrate. Abstracting an organism from its environment, ignoring this dialectic of interpenetration, is a reductionist step which methodology may demand but which will always mislead. The second lesson is that organisms are not passive responders to their environments. They actively choose to change them, and work to that end. The great metaphor of what Popper rightly called 'passive' Darwinism, natural selection, implies that organisms are the mere playthings of fate, sandwiched as it were between their genetic endowment and an environment over which they have no control, which is constantly setting their genes and gene products challenges which they can either pass or fail. Organisms, however, are far from passive: they – not just we humans, but all other living forms as well – are active players in their own futures.

BEING AND BECOMING

The first phases of the life cycle are those of development – ontogeny. From the moment of fertilization, cells grow, divide and hence multiply. Daughter cells begin to align themselves with respect to one another, to migrate to specific regions within the developing embryo. Within each cell, particular genes are switched on, and others off, in intricate sequences, as originally totipotent cells become specialized and the mature form of the organism unrolls from its undifferentiated state. Development poses a particular problem for living organisms, one which is quite distinct from that which we humans face in constructing artefacts. Consider the assembly line on which a car is built. Raw materials – sheet metal, plastic, glass – come in at one end. Engine blocks are cast, panels are beaten and fixed in place, and almost before our eyes a vehicle is assembled, checked and released, ready to be driven off. But it is only at the very end that a fully formed, functional car appears. No one imagines that at its halfway stage of assembly the car will function in miniature, so to speak, able to be driven at half-speed, or to carry two instead of four passengers.

Living organisms are quite different. From very early on in their development they have to be capable simultaneously of a quasi-independent existence, and of growing further towards maturity. Moreover, the attributes that enable them at any one moment to maintain their existence are not always merely 'miniature' forms of those they will need in adulthood. This is obviously true for some life forms. Frogs' eggs become tadpoles become frogs; butterflies' eggs become caterpillars and chrysalises before butterflies emerge. Each stage requires a radical transformation of body plan, yet during each transformation the functions necessary to life must be preserved. But it is also true in quite subtle ways for organisms which seem to show linear developmental trajectories without such radical breaks. When a newborn baby suckles at its mother's breast, the suckling reflex is not simply an undeveloped form of the chewing technique that will be needed when the child switches to solid food; quite different neural and mechanical processes are involved. Life demands of all its forms the ability simultaneously to *be* and to *become*.

Dichotomously genetical thinking wishes always to partition – first splitting 'nature' from 'nurture', and then adding them together again. So both being and becoming are regarded as the products of the additive effects of genes – nature – and 'environment' – nurture. By now it should be clear that I regard this dichotomy as spurious. The unrolling processes of development are best understood in terms of a different dichotomy, that between *specificity* and *plasticity*. One can consider these terms as an extension of Dobzhansky's concept of norm of reaction, which I introduced at the end of the previous chapter. Many ontogenetic processes are relatively unmodifiable by experience. For example, we humans, like some but not all other mammals, are born with our eyes open, already able to focus them reasonably well and to see and perceive colours, shapes and movements. This means that the pattern of connections by which the light-sensitive cells of the retina of the eye connect to the brain through the optic nerve must already be well established. During the first few years of life, both eyes and brain grow, though not in proportion to each other. Because of this growth, the actual physical chain of connections between retinal cells and brain neurons – the synaptic junctions between them – cannot remain the same. As both eyes and brain grow and mature, the connections must be broken and reformed many times, yet the overall pattern of the relationship between eye and brain must be maintained if vision is not to be impaired. That is, at any one moment eye and brain must both be adapted to current needs and must also be in the process of changing to meet future ones – both to be and to become. Furthermore, for vision to continue normally – that is, for functional specificity to be retained – this process must be relatively impervious to experience, to environmental contingency. But not entirely so. It is possible for the pattern of connections to be modified, at least during certain critical periods of development. For instance, rearing cats in environments of horizontal or vertical stripes, or with only one eye open, lastingly changes the patterns of synaptic connectivity.[2] (Such experiments also enabled methods to be developed to correct the visual defects in humans born with a squint, who would otherwise lack effective binocular vision.) This, then, is the measure of plasticity – the norm of reaction – which can be imposed upon developmental specificity. But both specificity and plasticity are embedded properties of the organism; both, if you like, are completely made possible by the genes, and completely made possible by the environment. They cannot be partitioned.

INSTRUCTION, SELECTION, CONSTRUCTION

Two contrasting metaphors have been used to describe the process by which multicellular organisms are constructed, both deriving from the language of information theory and both applied originally to the immune system's capacity to respond to the effectively infinite variety of challenges the environment might throw at it: *instruction* and *selection*. Confronted with invasion by a foreign organism or toxic substance (an *antigen*), the immune system rapidly synthesizes seemingly tailor-made proteins – *antibodies* – which can stick to the surface of the invading cell or bind to the antigen, thus labelling it so that it can be destroyed. How is this done? Immune systems which evolved to protect from microorganisms have no way of predicting in advance which molecules might confront them, certainly not the myriad of industrial chemicals which now pollute our environment, and which did not exist during human evolution. Yet the system is capable of making antibodies to counter a seemingly indefinitely large number of wholly novel substances.

Back in the 1960s, there seemed to be two alternative possibilities. On the one hand, the antibody-producing cells might simply be utilizing general-purpose mechanisms capable of making antibodies of any required shape. The arrival of an antigen would act as an instruction to the cell as to the shape of the protein required to stick to and hence immobilize the instructor. Alternatively, there might exist among the population of potential antibody-producing cells a wide variety of types already roughly tailored, one or more of which would be likely to fit at least approximately any potential antigen. The arrival of the antigen would trigger a massive expansion of production of the cells producing the antibody that best fitted it, and where appropriate the final adjustments to the antibody protein to make the fit more perfect. This is the selectionist model. The difference is between instruction as bespoke tailoring and selection as off-the-peg purchasing.

Despite an initial prejudice in favour of instruction as apparently the more obvious of the two mechanisms, the evidence soon proved convincingly that the immune mechanism works by selection. Gerald Edelman, who shared a 1972 Nobel prize for his immunological research, subsequently expanded the selection theory into a general model to account for ontogenetic processes, applying it particularly to the development and 'wiring' of the brain. He has called the mechanism 'neural Darwinism'.[3] I don't like the term (nor does Francis Crick, who has called it disparagingly 'neural Edelmanism'[4]), as the process Edelman describes is neither homologous nor adequately analogous to Darwinian natural selection. Neural Darwinism is a seductively misleading metaphor, but the concept embedded within it is non-trivial and important to grasp. Although the issues it raises are most significant for the human nervous system, the body's most complex structure (perhaps the most complex in the living world – or even, as some have suggested, in the entire universe), the principles of neural Edelmanism apply to development more generally.

To appreciate the problem, consider the adult human brain. Its 1.5-kilogram mass contains up to a hundred billion neurons and ten times as many supporting cells, known as glia, surrounding them. This cell mass is highly structured. It is divided into numerous functionally specialized regions, and in each region the cells are arranged in a highly ordered pattern. Thus the surface of the brain is formed by a thin highly convoluted 'skin', about 4 millimetres thick – the cerebral cortex (grey matter). The cortex is packed with neurons arranged like a layer cake in six 'strata'; the pattern is readily observable in side view through an optical microscope if the cells are appropriately stained. Less well observable is the fact that, as viewed from the cortex's upper surface, the cells are also organized into an array of functionally distinct columns. Closer examination of neurons reveals that they also show a number of distinct shapes, resembling pyramids, stars, baskets, and so on (Figure 6.1). As if this were not enough, each neuron is connected to others, some its neighbours, some distant from it, by fine fibres that radiate out from the cell body. Some of the fibres (dendrites) collect incoming signals, and at least one of them (the axon) transmits the information carried by these signals onwards to the other neurons, making contact with their dendrites by way of junctions called synapses. Any one neuron may carry up to a hundred thousand of these synaptic connections. Some of the connections are internal to the brain, enabling neurons to communicate with their colleagues. Others – such as the great cable of axons that runs from the eye down the optic nerve, first to a region deep inside the brain called the lateral geniculate, and from there to the 'visual' regions of the cortex – carry inputs from the external world. Still other nerve tracts lead out from the brain, connecting via the spinal cord with the body's musculature and internal organs.

This enormously complex structure must be generated within nine months of the moment of fertilization, so as to be largely functional by the time of birth. Of course, there's a lot of post-natal development still to go. Many of the glial cells are not yet in place at birth. And, even more important for the functioning of the brain, synapses are still sparse at birth. During the next few years of development, no fewer than 30,000 synapses a second will be created under each square centimetre of cortex, until the full complement of a hundred trillion (10^{14}) are present and functioning. To put that number into perspective, it is about 20,000 times more than the entire human population of the planet.

But even to get to the stage of the brain at birth requires the creation of about a million cells an hour, day in day out, throughout the entire gestation period – a formidable enough challenge if the brain were simply growing smoothly, like a steadily inflating balloon. But it isn't. The first observable step is taken when the embryo is no more than eighteen days old and 1.5 millimetres long, when the hollow ball of cells that constitutes the gastrula develops a groove along its surface, thickened and enlarged at the forward end, which will in due course become the brain. As development proceeds the groove deepens and

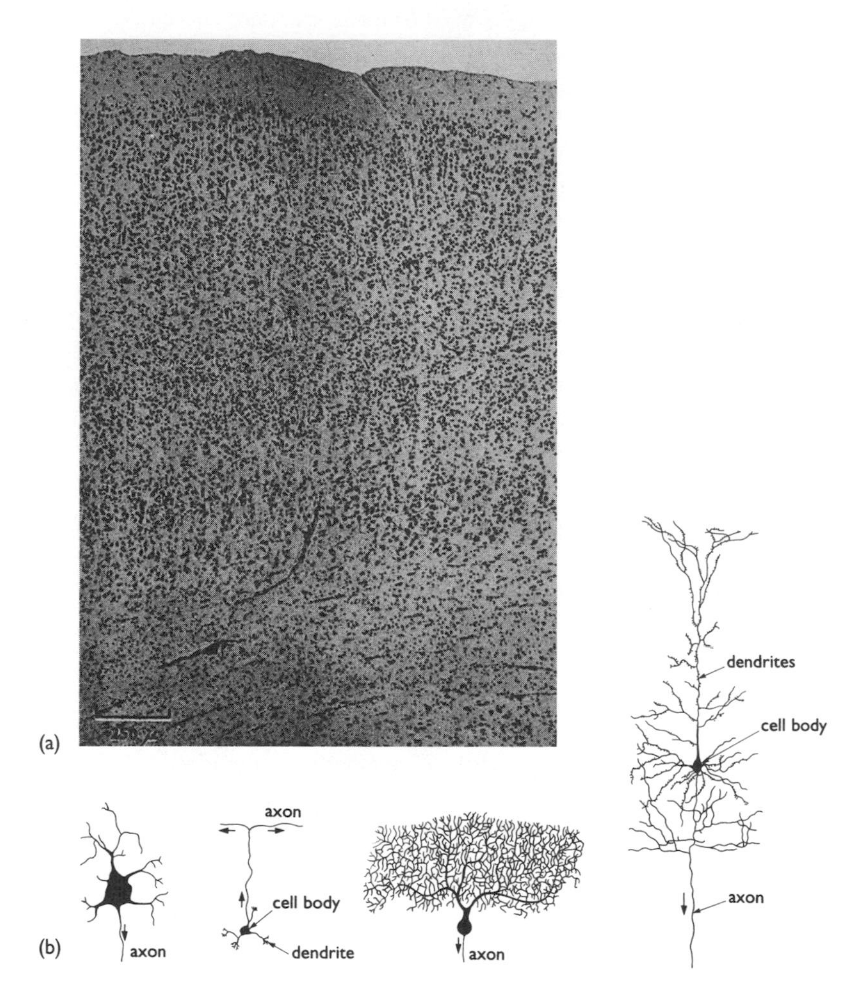

Figure 6.1 (a) A cross-section through the cortex of the human brain. Each black dot is a single neuron. (b) Some of the many varieties of shapes of neurons, each specialized for a different function.

its walls rise higher, move towards one another, touch and seal over; the groove has become the *neural tube*. By twenty-five days, when the embryo is about 5 millimetres in length, the tube begins to sink below the surface of the embryo. Its central cavity will become the central canal of the spinal cord and form the fluid-filled spaces within the brain itself (the ventricles). The head end of the tube begins to swell, and to show the beginnings of the three major divisions of fore, mid- and hind-brain (Figure 6.2).

In the next few months of embryonic development, precursor cells to all the billions of neurons and glia which will ultimately constitute

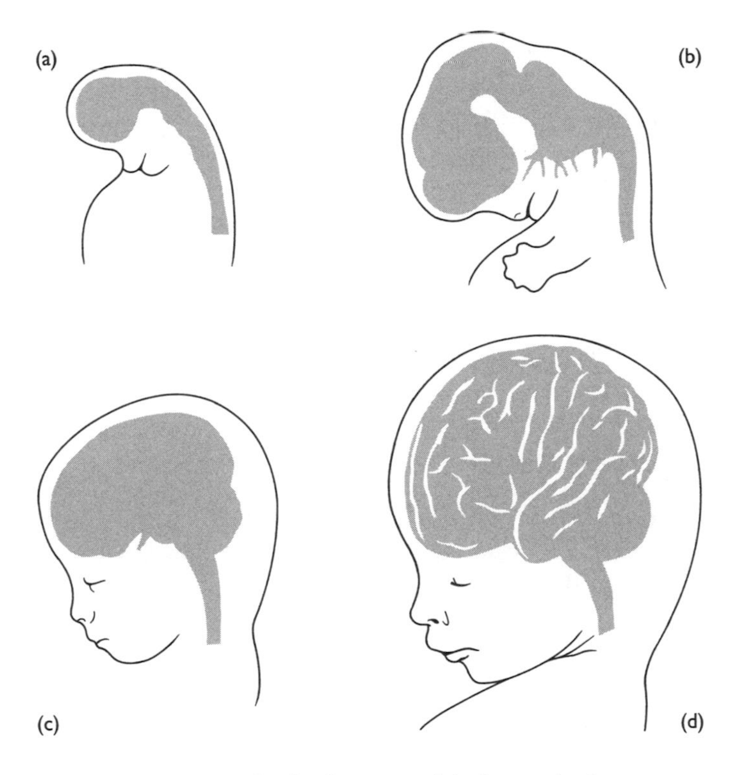

Figure 6.2 The development of the human brain: (a) 3-week embryo, (b) 7-week embryo, (c) 4-month foetus, (d) newborn infant.

the brain begin to separate from the neural tube. The precursor cells are thus not formed in the developing brain at the sites at which they will end up as mature neurons and glia, but in the vicinity of the neural tube and ventricles. They are then required to migrate from their places of origin to their ultimate locations, distances which may be tens of thousands of times their own length – equivalent to a human navigating a distance of 20 kilometres. How do they find their way? Does each cell know where it is going, and what it is to become before it arrives at its final destination? Is it equipped with a route map, or is it, as in the instructional model of the immune system, a general-purpose cell which can take on any appropriate form or function, depending on its final address within the brain?

To many of these vital questions there are still no complete answers; as I pointed out in the previous chapter, the great expansion of genetic knowledge in recent decades has yet to be matched by a comparable increase in the understanding of development. But several mechanisms are known to play a part. In the developing brain it is the glial cells that begin the migratory pattern. As they move away from their sites of origin and towards what will become the cortex, they spin out long tails, up which the neurons can in due course climb. As Edelman and others have shown, the cell membranes of both neurons and glia contain a particular class of proteins called *cell adhesion molecules* (CAMs). In the developing tissue the CAM molecules work rather like crampons: they stick out from the surface of the membrane and cling to the matching CAM on a nearby cell. The neurons are thus able to clutch the glia and ratchet themselves along their tails (Figure 6.3). As a further trick, the migrating cells also lay down a sort of slime trail of molecules related to the CAMs – *substrate adhesion molecules* (SAMs) – which provide additional guidance for the cells following along behind.

But what provides the map references for such cellular route marches? Both distant and local signals must be involved. One way of signalling direction is to have already in place some target cell or tissue towards which the migration can be directed. Suppose the target is constantly secreting a signalling molecule, which then diffuses away from it. This will create a concentration gradient, highest at the target

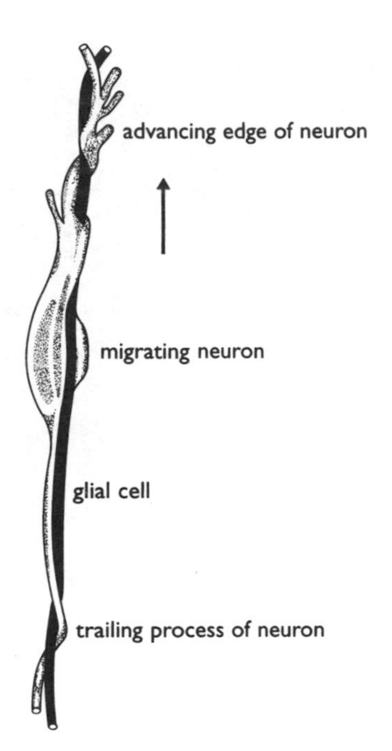

Figure 6.3 A migrating neuron climbs up a glial cell fibre (in black).

and progressively weaker at increasing distances. If the migrating cell can sense the signalling molecule and move towards it, rather as bacteria can swim towards sources of food, then it will eventually arrive at the target. In the 1950s Rita Levi Montalcini identified one such signalling (or *trophic*) molecule, which she called nerve growth factor; by the time she was awarded her Nobel prize for the discovery, in 1986,[5] it had been recognized as but one of a growing family of such molecules (Figure 6.4).

Trophic factors can provide the long-range guidance which enables the growing axons of motor nerves to reach out and find their target muscles, or the axons from the retinal neurons which form the optic nerve to track their way to their first staging post within the brain, the lateral geniculate. However, the migrating cells or growing axons also need to keep in step with one another – each has to know who its neighbours are. The diffusion of a local gradient molecule, together

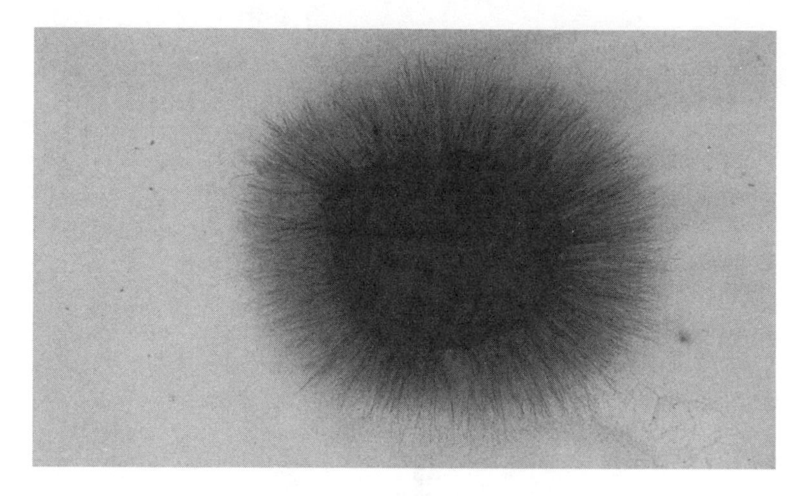

Figure 6.4 The outgrowth of nerve fibres from a secondary ganglion (cluster of neurons) treated with nerve growth factor.

with the presence of some types of chemosensor on the axon surface, could enable each to determine whether it has neighbours to its right and left and to maintain step with them (Figure 6.5).[6] The entire troop of axons would then arrive in formation at the lateral geniculate and make appropriate synaptic connections, thus creating in the geniculate a map – albeit a topographically transformed one – of the retina, rather like the relationship between the London Underground or New York subway system and the maps of them on display at stations. Indeed, the brain holds many such maps, multiple maps for each of its input sensory systems and output motor systems, maps whose topology must be preserved during development.[7]

The process just described would be compatible with an instructionist model. Each axon is kept on course by instructions from its environment, both the trophic factor diffusing from the target region and the relationships with its nearest neighbours directing it to its final site. There is some evidence that a considerable part of the nervous system's development can be accounted for in such a model.[8] However, Edelman drew attention to another vital feature of develop-

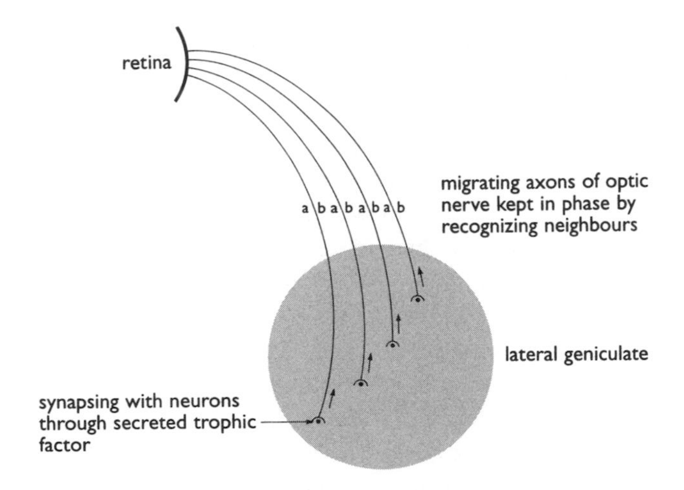

Figure 6.5 Local and distant guidance of migrating axons of the optic nerve. a and b are local recognition signals keeping the axons in step.

ment. During embryonic development there is a vast overproduction of cells: many more neurons are born than subsequently survive. Since more axons arrive at their destination than there are target cells to receive them, they must, argues Edelman, compete for targets. Those that do not find them wither away and die. The argument actually goes further: it is not only neurons and their axons which are overproduced, but the synapses too. There is a superabundance of synaptic production, a veritable efflorescence. But if synapses cannot make the appropriate functional connections with the dendrites of the neurons they approach, they too become pruned away and disappear. In this model of development, because there is competition for scarce resources – trophic factor, target cell, synaptic space – there is also selection. And now we have only to imagine that it is in some way the 'fittest' of the neurons and synapses that win out in the competition, and we arrive at Edelman's 'neural Darwinism'.

Selection in this sense can account for local but not distant processes. Long-range order, the migration of cells and the growth of axons over long distances, would seem to require something more – the execution of some internal programmes of both individual cells and the collectivity of cells acting in concert. Even though synapses from only one particular neuron may end up making successful connections with its target cell, if the others had not been present during the long period of growth and migration it is doubtful whether a single nerve axon would have been able even to reach the target. The survival of one depends on the presence of the many. Overproduction and subsequent pruning of neurons and synapses may at one level of magnification look like competition and selection; viewed on the larger scale, they appear as cooperative processes.

As a comparable example, it takes only one sperm to fertilize an ovum. In the vulgarly macho language that one has come to expect from some popular writers about biology, combining ultra-Darwinist rhetoric with sexual prurience, this 'fittest' successful sperm is often interpreted as being the 'winner' of a competition amongst the many hundreds of millions in an ejaculate.[9] In fertilization, the head of the sperm cell – containing the nucleus – fuses with the egg. Yet introduce just this single 'fittest' sperm into the vagina and the chance of it surviving to reach and fertilize the ovum are minuscule; a high sperm count improves fertility, helping more sperms to survive their passage through the vagina, even though only one will ultimately enter the ovum and complete the fertilization. The single 'fittest' sperm must in fact cooperate rather than compete with the rest if fertilization is to occur at all. (Furthermore, it is increasingly apparent that the ovum is not merely the passive recipient of the victorious sperm, but plays an active part in the process. Fusion requires the sperm's enzymes to be activated by secretions from the female reproductive tract, and sometimes also by the protrusion from the egg's surface of small membranous 'fingers' that draw the sperm into the egg.[10])

Instructive and selective mechanisms are thus only part of the picture of development. The maintenance of stability requires the entire ensemble of cells to cooperate, to act collectively. In a non-trivial way, each depends on the others in the creation and preservation of the dynamic pattern of connections which maps the world onto the sense organs, the sense organs onto the brain, and then, via the brain and the musculature, imposes new patterns on the world beyond. This is why I want to argue that we need to transcend both instructionist and selectionist metaphors. Development is essentially a constructivist process;[11] the developing organism, in its being and its becoming, in its specificity and its plasticity, constructs its own future.

CHANCE AND DETERMINISM

Even the constructivist model of development discussed above implies a degree of determinism, albeit in this case a richer concept than that of the unidimensional gene. But we need to go beyond this in emphasizing the role of chance, of contingency, at all levels of analysis of living systems. Consider the micro-level of the individual cell and its subcellular components. Biochemists deal of course not with individual cells or with individual copies of their molecules, but with aggregates of millions, and on this scale properties become fairly predictable. But what is predictable for the mass does not apply to the individual. The role of mitochondria, for instance, in carrying out an exquisitely controlled series of reactions by which the products of glucose breakdown are oxidized and ATP is synthesized from ADP, has been minutely studied, and the reactions are known to depend on precise fluxes of hydrogen ions across the mitochondrial membrane. Yet if one considers an individual mitochondrion at the normal pH (the degree of acidity or alkalinity) of the cell, there are likely to be only a matter of thirty or so free hydrogen ions available within it – a number so small that fluctuations due to thermal noise make it quite impossible to calculate the distributions of the ions precisely. Chance at this level affects all cellular processes, including, as has long been recognized, the random mutations in DNA structure induced by cosmic radiation or other mutagenic agents.

Similar considerations apply to the role of chance in development. Lewontin has pointed out that even in *Drosophila*, which is supposedly bilaterally symmetrical, as a result of random developmental events the number of bristles on a leg on one side of the body may not match the number on the opposing leg. And what is true for the role of contingency in the development of *Drosophila* is certainly true for humans. For instance, identical twins share identical DNA, yet from the moment of conception and cell division the relative locations of the two embryos to the placenta and to the environment of the uterus affect their development in chance ways. Developmental divergence increases with every cell division, and after birth with every random experience of each twin. If contingency features in the factors shaping the development of any individual organism, still more does it apply when we come to consider the role of chance and random events in evolutionary processes, as will become apparent in the chapters that follow. Chaos theory has made much play of the butterfly effect in modelling the weather, though an old saying put the case much more simply when it described how, for the want of a nail, the shoe, the horse, the messenger and ultimately the battle is lost. It is just this combination of predictability and unpredictability that distinguishes living systems and processes from the much simpler events that form the terrain of the sciences of physics and chemistry.

HOMEOSTASIS AND HOMEODYNAMICS

Claude Bernard's slogan – the constancy of the internal milieu, the internal environment of multicellular organisms – has become one of the central organizing themes of physiology. The inevitable fluctuations in the world outside the organism, of temperature, for example, or of available foodstuffs, are damped, compensated for, so as to preserve this constancy. Increases in external temperature provoke sweating, and decreases result in restricting bloodflow to the surface skin, so as to maintain the internal temperature more or less constant at (in humans and other mammals) around 37.5°C. Food deprivation, which lowers blood glucose levels, mobilizes the sugar stored in the form of glycogen in the liver, or stimulates the breakdown of fat. It also results in changes in the behaviour of the organism: hunger induces food-seeking behaviour in all of us. So too with many other features of the internal environment, from the pH of the cell, kept just on the alkaline side at 7.4, to the balance between sodium and potassium ions or the ratio of ATP to ADP (adenosine diphosphate) within the body's cells.

The level at which any of these variables is maintained is called its *set point*. It is the stability of the set point that is implied in the term homeostasis, and introductory physiology textbooks treat at some length the mechanisms that maintain such stability. The metaphor which is used to illustrate it is frequently that of the thermostat in a house's central heating system. The thermostat's temperature control is set such that if the temperature falls below its set point, the heating system comes on and the temperature rises; as it increases above the set point, the system is switched off. The result is that if you were actually to record the temperature in the room under thermostatic control, it would not be precisely constant, but would oscillate slowly around its set point. How fast and how far the oscillations occur depends on the sensitivity of the thermostat and the efficiency of the heating system: if the thermostat is not particularly sensitive the oscillations may be uncomfortably large; if it is too sensitive it will switch on and off so rapidly that the heating system may break down. Stability is best achieved not by attempting to keep the temperature perfectly constant, but by designing for optimal frequency and size of the oscillations around the set point. The type of oscillations that such thermostatic homeostasis imply are shown in Figure 6.6(a).

That is as far as the biological metaphor is usually taken, but let's pursue it a little further. In practice, even for rooms whose temperature is controlled by a central heating system, this description of regular oscillations around a fixed set point is inadequate. Most domestic heating systems are programmed not to provide an even temperature day and night, but to run at a lower temperature or switch off entirely at night, and indeed, if no one is home during the day, to do so during the middle period of the day as well. So the actual pattern of temperature variation in rooms controlled by such a thermostat is more likely to be as shown in Figure 6.6(b); that is, to show a diurnal rhythmicity. Thus there is a super-rhythm imposed upon the homeostatic oscillations, like the coiled filament of a tungsten lamp. More sophisticated thermostats can be programmed on a seven-day cycle, recognizing that many of us have different patterns of residence

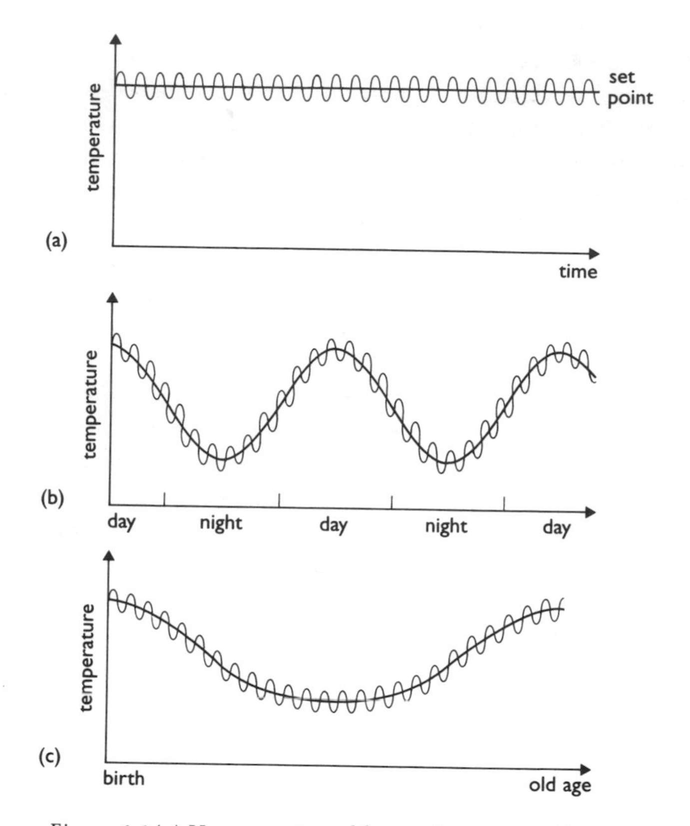

Figure 6.6 (a) Homeostatic and homeodynamic oscillations; (b) diurnal rhythmicity; (c) life-cycle rhythmicity.

and temperature needs at weekends and on weekdays. There are longer-term fluctuations too, if the system is shut off in summer or during holidays. And viewed over a lifetime, one might want the mean temperature greater during a time when there are infants or elderly people in the house (if they can afford the fuel bills) than when it is inhabited by adults in their middle years. A lifetime's view of thermostatic settings might therefore look something like Figure 6.6(c).

So, viewed on a longer time-scale, even a room thermostat does not display homeostasis in the sense of 'staying the same', but incorporates a range of cycles and 'hypercycles'. Homeostasis is replaced by homeodynamics. What is true for this simple mechanical metaphor is true to an even more dramatic extent in living organisms. Seeing them as merely homeostatic is to deny them lifelines, to fall into the empty-organism trap that the gene's-eye view of the world demands. The set points around which the moment-by-moment fluctuations in an individual's biochemistry oscillate on the microscale themselves change during the trajectory of a lifetime. Our body temperature, steroid hormone levels and neurotransmitter levels maintain diurnal rhythms. Some 52 per cent of the human population aged between about 13 and 50 experience monthly hormonal cycles which significantly affect their patterns of life. The remaining 48 per cent may also show comparable changes, though hitherto researchers have scarcely bothered to look. Other monthly and annual cycles are only as yet dimly understood, from the rising sap of spring to the autumnal melancholy which may even result in seasonal affective disorder ('SAD') for some of us in the gloom of high-latitude winter. And every individual reading this book, just as is its author, is part of the way along the longest individual trajectory of all, which takes each of us from single fertilized cell, through the 10^{14} cells which constitute our adult existence, and ultimately to death.

Lifelines are thus inherently homeodynamic.[12] The present instant of our, or any organism's life, is simply inexplicable biologically if considered merely as a frozen moment of time, the mere sum, at that moment, of the differential expression of a hundred thousand genes. Each of our presents is shaped by and can only be understood by our pasts, our personal, unique, developmental history as an organism. Not for the first time in this book, and not for the last, I repeat my adaptation of Dobzhansky's famous statement: 'nothing in biology makes sense except in the light of history'.

Even the moment-to-moment stability of the organism is maintained not statically but dynamically. It would be an easy mistake to make to assume that life cycles mean a period of growth – say from conception to adulthood in humans – then a long period of relative stasis, and then finally a decline into old age and death. Even Shakespeare, with his seven ages of Man, knew it was more complex than that.

Long before molecular biology was dreamt of, biochemists had revealed what became known as 'the dynamic state of body constituents'. Each cell in the adult body has its own life cycle, from birth at mitosis to death and replacement within a few days, weeks or months. The exceptions are the neurons of the brain, which make up a non-dividing cell population and which are not replaced when they die; most therefore last us a lifetime. By contrast, the red haemoglobin-containing cells of the bloodstream live a mere 120 days before dying and being replaced.

The life and death of any cell proceeds on its course relatively independently of the life and death of the molecules of which it is composed. The complex macromolecules, the proteins, nucleic acids, polysaccharides and lipids within each cell have life cycles of their own, continually being broken down and replaced by other, more or less identical cells. The average lifetime of a protein molecule in the body of a mammal is around a fortnight. In an adult human, proteins constitute some 10 per cent of body weight, so some 24 grams of protein are being broken down and a fresh 24 grams synthesized every hour of every day – half a gram, or more than a billion billion molecules of protein a minute, throughout our adult life. Why this ceaseless flux? Why not build bodies like houses: constructed once, altered, maintained and repaired as necessary, but basically unchanging until finally demolished?

METABOLIC WEBS AND THE PRESERVATION OF ORDER

The answer is simple: just as a room thermostat demands oscillations in order to preserve stability, living systems need to be dynamic if they are to survive, able to adjust themselves to the fluctuations which, even in the best-buffered internal milieu, their cooperative existence as part of the greater unity of the organism demands. Frederick Gowland Hopkins understood this well when he composed the definition of life which forms the epigraph to this chapter, and which informed the way biochemistry was taught to generations of Cambridge undergraduates, myself included. Hopkins, one of the founders of modern biochemistry, and the discoverer of vitamins, among many other achievements, was a chemist by training, yet would never for a moment have considered that biochemistry could simply be reduced to chemistry. The school he founded in the first decades of this century was wedded to the concept of dynamic biochemistry, and it is to this irreducible dynamism as the generator of stable order that we must now turn in order to understand how, having constructed itself through the processes of development, the organism is able to preserve its integrity and act upon the external world. These are the phenomena of autopoiesis.

Biochemistry began in reductionist mode. Its precursors, the nineteenth- and early-twentieth-century organic and physiological chemists, took to themselves the task of analysis, of decomposing cells and organisms into their constituent molecules, small and large. Here was life – nothing more than organic chemistry. Chemically synthesized urea was identical to that excreted by the body; the mysterious 'protoplasm' and indeterminate 'colloids' which were supposed to constitute the stuff of life could be turned into purified and crystallized proteins. In due course these, and nucleic acids too, would be synthesized chemically.

So what breathes life into these complicated but no longer mysterious chemicals? First, they are constantly undergoing many complex reactions of synthesis and degradation, reactions whose precision is beyond the scope of mere human chemists. Furthermore, these reactions are taking place not as chemists would make them happen, by the use of strong reagents, acidity or alkalinity and extremes of temperature, but in the tranquillity of cells whose internal pH never varies greatly from neutrality, and whose temperature remains constant within a degree or so. The agents that catalyse such reactions are enzymes, and much of twentieth-century biochemical research has been concerned with purifying the thousands of individual enzymes each cell contains, and studying in isolation the chemistry of the reactions they bring about. Each enzyme works on a particular molecule (known as its *substrate*) and converts it to one or more products. Theoretically, enzyme reactions are all reversible, and if they are studied in isolation in a test-tube, eventually an equilibrium develops between the concentrations of substrates and products. The speed with which the enzyme works can be influenced by its environment – the presence of particular ions which may activate or inhibit it, the temperature, pH and so on – but the final balance point, the equilibrium between substrates and products, is unaffected. Such an enzyme-catalysed reaction may be written as

$$A + B \underset{k_2}{\overset{k_1}{\rightleftharpoons}} C + D \qquad (1)$$

which represents the conversion by a reaction of substances A and B into substances C and D. The equation is reversible, which means that depending on the conditions it can proceed forwards, from left to right, or backwards. Which direction it proceeds in depends on the so-called rate constants for the forward and backward reactions, shown in the equation as k_1 and k_2. (If, like me, you hate equations and find these algebraic representations hard to follow, don't worry – we'll be out of them before long, and all you need to follow the argument is the general idea, not the details.)

The second crucial aspect of living systems is that, even when catalysed by an enzyme, many reactions – for example those involved in the synthesis of proteins or nucleic acids – require an input of energy. Thus cells need energy to sustain themselves even before they begin to act upon their surroundings. Muscles contract, nerve cells send messages, endocrine cells produce hormones, and so on. The original source of such energy for virtually all living organisms is the Sun. Green plants trap the Sun's light energy by way of photosynthesis and use it to convert atmospheric carbon dioxide and water into sugars through a complex series of reactions whose study has provided joy and frustration alternately to several generations of biochemists, but which is now pretty well understood. Other life forms can then in turn burn the sugars made by the plants, via a series of controlled enzyme-catalysed reactions, in order to release the energy trapped in the sugar molecules in a form they can use. Pivotal to this process is ATP, introduced in Chapter 2 as the 'energy currency' of the cell. ATP is synthesized as glucose and other sugars are burned, and broken down again (to ADP) to release the energy for both self-maintenance and action.

The reductionist approach to the chemical dynamics of life was therefore to disassemble cells into their constituent molecules, and follow each individual enzyme reaction through which they are transformed in terms of both its chemistry and its energetics. Thus the enzymes which catalyse the breakdown of glucose and release its energy are coupled to others which use the energy to synthesize ATP from its precursor (ADP); this breaking-down is called *catabolism*. Reciprocally, synthetic reactions, such as those which build proteins from their constituent amino acids, require ATP, breaking it down to ADP in the process; this is *anabolism* (Figure 6.7). By the 1930s, when these mechanisms began to be deciphered, chemistry had spent a hundred and fifty years – since the days of Lavoisier – dealing with the energetics of such reactions, studied within the framework of the science of thermodynamics. Thermodynamics is concerned with equilibria, the final balance points between energy-yielding and energy-providing reactions, and the mathematics and physics of such equilibria were well understood. Simplistically, the net effect of all the energy-utilizing and energy-generating reactions should be that the

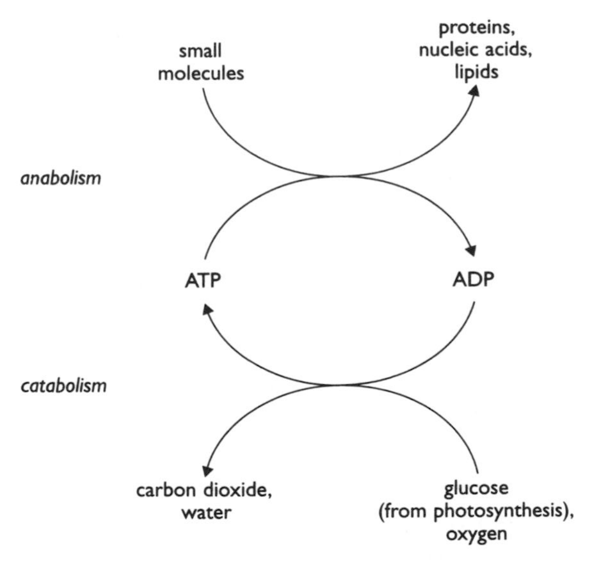

Figure 6.7 The catabolism/anabolism cycle.

cell is in thermodynamic and catalytic balance, and should equate to the life-process itself. Indeed, during the 1920s and 1930s, physiologists and biochemists spent much time devising complex accounting experiments, measuring the calorific value of the food intake and the excreted waste and energy output of living organisms – from plants to humans – kept in closed metabolic chambers, to prove that this was the case. Healthy organisms are, energetically speaking, in balance.

While this is of course true (if it were otherwise, the implication would be that life was violating key physical principles), if we are to interpret the complexity of the processes occurring within living systems, then we have to take them out of their closed metabolic cages. And it is just at this point that the reductionist approach, brilliant at the analysis of individual reactions, begins to come apart. Equilibrium mathematics, whether for chemical reactions or thermodynamics, deals with closed systems. For the experiments or the formalisms of the maths to work, they have to start with a given quantity of initial components and a given input of energy in the form of heat or whatever. They are then sealed off from the rest of the universe and allowed to run to completion, until the reactions have stopped or come to some balance point which can be calculated based on the rates of forward and backward conversion in equations like (1) above. But living systems are not sealed off in this way: they are open, as we have seen, and in constant interchange with their environment. Raw materials – glucose, oxygen, other small molecules and ions – enter the cell, while waste molecules and other exports leave. Life is characterized not by the static balance of completed reactions, but by dynamic equilibrium. This is the first component of Hopkins' definition, in which stability results from the constant flux of components and their reactions – traffic in and out of the cell. Formulations like equation (1) describe test-tube isolates, not real-life phenomena.

The thousands of chemical reactions taking place at any moment within the cell constitute a complex interacting web. Having studied each individually, the logical approach of the reductionist is to attempt to build them up into sequential chains, recognizing that the products of one enzyme-catalysed reaction will immediately serve as the substrates for another. For example, when glucose is broken down, ultimately to be oxidized to carbon dioxide and water, the initial eight reaction steps, each catalysed by an individual enzyme, result in the 6-carbon glucose molecule being converted into two 3-carbon molecules of pyruvic acid, with the simultaneous synthesis of a number of ATP molecules. One can write such a reaction sequence abstractly as the conversion of a mythical substance W to a final product Z by way of three enzymes and two intermediates:

$$W \underset{W-ase}{\overset{1}{\rightleftharpoons}} X \underset{X-ase}{\overset{10}{\rightleftharpoons}} Y \underset{Y-ase}{\overset{100}{\rightleftharpoons}} Z \qquad (2)$$

Each reaction has a characteristic set of rate constants, given here in arbitrary units for the forward direction in the equation. The overall rate at which Z is produced will be governed by the slowest of the reactions in the chain – the so-called *rate-limiting* reaction – in this case the enzyme W-ase. In practice it often turns out that the rate-limiting step is one of the first in the sequence – obviously advantageous so far as the cellular economy is concerned. But in addition, because the rates of enzyme reactions are greatly affected by factors such as acidity and ion concentration, the enzyme W-ase can serve as an effective control point for the entire sequence. Suppose, for instance, that the final reaction in the sequence, catalysed by Y-ase, produces not merely Z but also hydrogen ions (H^+), which increase the acidity of the solution, and that the increasing acidity slows down W-ase. The result will be that the end-product of the reaction, Z, regulates the rate of its own production by feedback *inhibition* of W-ase. The reaction sequence is thus a self-regulating one:

$$W \rightleftharpoons X \rightleftharpoons Y \rightleftharpoons Z + H^+ \qquad (3)$$

I should now confess that I have drawn this example, almost verbatim, from the first book I ever wrote, *The Chemistry of Life*,[13] and it has appeared, without major qualification, in every edition from the first in 1966 to the most recent in 1991. And it is, regrettably, far too simple, framed by a reductionist mode of thought which I still only partially transcend. It is too simple because, of course, just as within a living cell as opposed to a test-tube one cannot abstract an individual enzyme reaction from the metabolic dance of the molecules, so one cannot abstract any single reaction pathway. Year after year, ever since I first qualified as a baby biochemist, one leading producer of biochemicals has issued a chart describing the metabolic pathways known to occur within a 'typical' mammalian cell – which probably means a liver cell. Just a small subset of these pathways is shown schematically in Fig 6.8. Even this is grossly oversimplified, because it is almost impossible to show reactions which occur within the four dimensions of space and time within which the cell exists by means of a two-dimensional representation. What it implies is that many of the substances represented by the mythic W's, X's, Y's and Z's of equations (2) and (3) participate not in one but in many interacting pathways, and the factors which may influence the rate of any individual enzyme reaction then multiply dramatically.

The implications of such interconnections are quite striking. Think of a piece of weaving, made up of threads of different colours. The weaving has a pattern, which resides not in any of the individual threads which constitute the warp and weft of the fabric, but in the product of their interactions. Furthermore, although the threads are individually quite weak, woven together they have considerable strength. And perhaps even more relevant, neither the pattern nor the strength depends on any one 'master thread'. Remove any individual thread and the pattern, strength and stability of the fabric are only marginally affected. It is like this with the metabolic web within every cell: once it reaches a sufficient degree of complexity, it becomes strong, stable and capable of resisting change; the stability no longer resides in the individual components, the enzymes, their substrates and products, but in the web itself. The more interconnections, the greater the stability and the less the dependence on any one individual component (a property called 'graceful degradation' by computer modellers).

The formal mathematical proof, due originally to the biochemical geneticist Henry Kacser,[14] who referred to the process as 'molecular democracy', is beyond the scope of this chapter; the analogy must suffice. As Kacser emphasizes:

There are therefore not simply two classes, 'controlling' and 'non-controlling' enzymes, but control is shared amongst all the enzymes . . . Descriptions

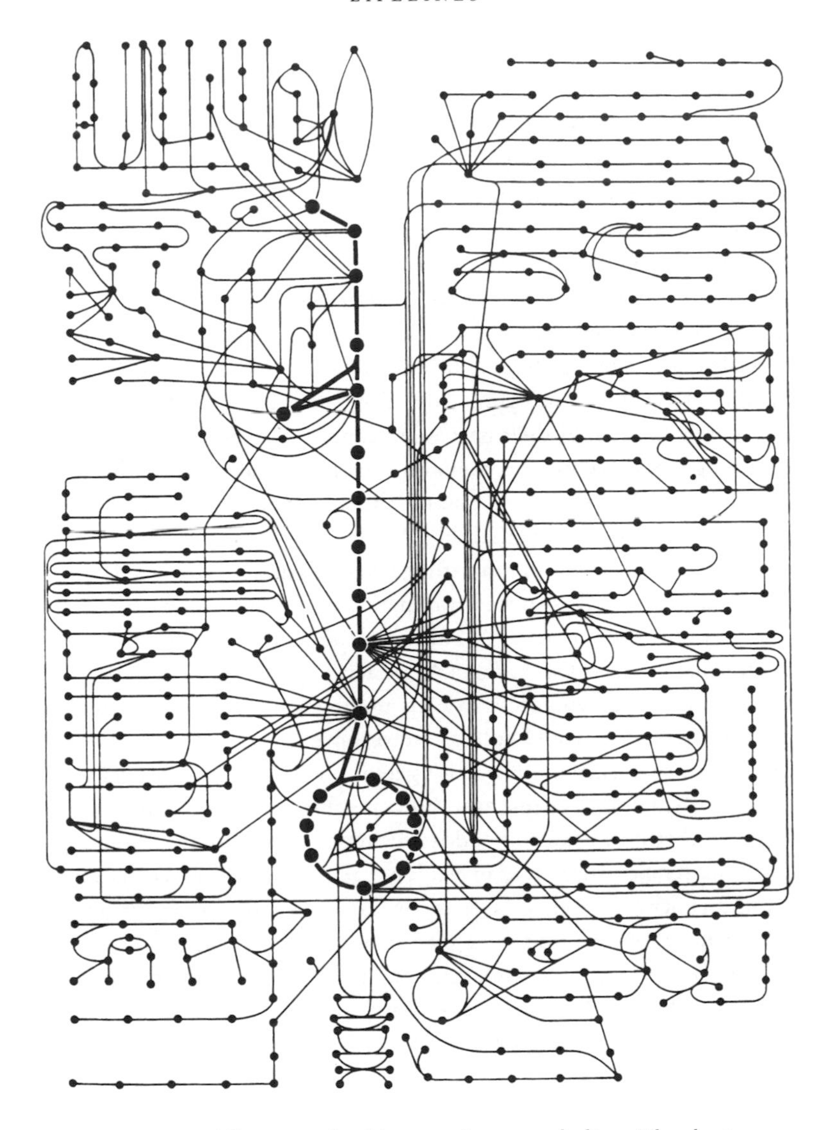

Figure 6.8 The network of intermediary metabolism. The chart shows about 700 small molecules interacting; each dot is a metabolite, each line a reaction pathway.

of enzymes as 'Pacemakers' or 'Rate-limiters' falsely introduces [sic] a classificatory concept where we are dealing with a continuum of values.

Mea culpa! But the metabolic web has a further advantage over one made of mere fabric. Unlike living systems, human artefacts such as fabric cannot compensate for the loss of any individual thread. The cellular web, however, has a degree of flexibility which permits it to reorganize itself in response to injury or damage. Self-organization and self-repair are its essential autopoietic properties. These properties of stability and self-organization, which Stuart Kauffman has described as 'order for free',[15] are the key to appreciating the fundamental irreducibility of living cells. Their metabolic organization is not merely the sum of their parts, and cannot be predicted simply by summing every enzyme reaction and substrate concentration that we can measure. For us to understand them, we have to consider the functioning of the entire ensemble.

But stability and self-organization also explain why the equilibrium achieved by the cell is indeed a dynamic and not a static one. The essence of the stability of the whole is that the individual components are in constant flux. Freeze them in reductionist immobility, and, like a skater on thin ice who needs to keep moving to avoid falling through, the cellular edifice would collapse into those individual components that we biochemists have for so long lovingly studied in dissected and impoverished isolation. Just as stability is achieved by a central heating system's thermostat not by endeavouring to maintain an absolutely constant temperature, but instead by accommodating oscillations around a fluctuating set point, so too in the cell. Studies of the dynamics of cell metabolism, pioneered over many years by Benno Hess in Heidelberg, have shown that levels of many metabolites and metabolic sequences display rhythmic oscillations, from the breakdown of glucose by glycolysis to the reproductive cycle of DNA synthesis, mitosis and cell division. Recently, new imaging techniques have shown too that intracellular messages, carried by the ubiquitous signals provided by the calcium ion, are also propagated as waves pulsing through living cells (Figure 6.9). In the open system of the cell, with a flow of energy passing through it and continual deviations from thermodynamic equilibrium, choreography is all.[16]

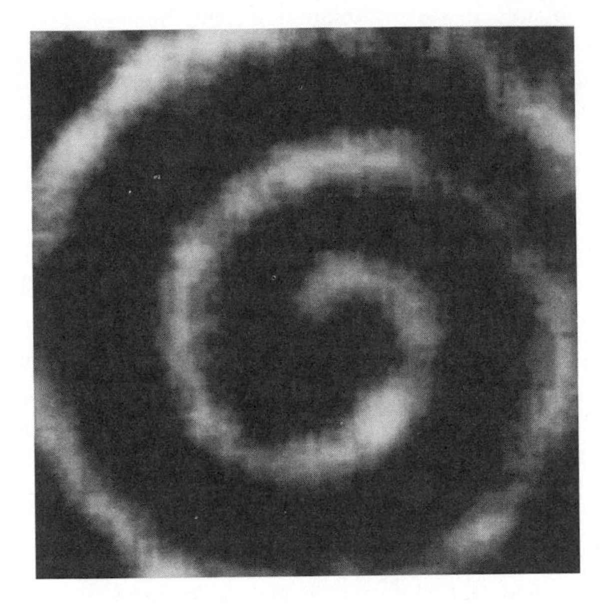

Figure 6.9 Spiral Ca^{2+} wave propagated in an oocyte.

STRUCTURE AND SELF-ORGANIZATION

Confine an appropriate number of substrates and enzymes, together with necessary energy sources, within a semipermeable membrane like that of the cell, and with a fair range of tolerance it is predictable that stable metabolic webs will emerge (I will leave to Chapter 9 theories of how such systems may have evolved). But cells are not simply bags containing semi-random mixes. Even gutted and dehydrated and pinned down on an electron-microscopist's grid, they reveal rich internal structures. Each (eukaryotic) cell has a nucleus, many mitochondria, photosynthesizing chloroplasts (if it is from a green plant), numerous small vesicles, and complex networks of internal membranes studded with tiny particles forming elegant rosette-like patterns, visible in the electron micrograph shown in Figure 3.3 (page 61).

It is possible to use techniques of centrifugation (discussed in Chapter 3) to separate these individual substructures of the cell, and it turns out that each has a specialized biochemistry. Chromosomes, and most of the cell's DNA, are in the nucleus. The rosettes which stud the internal membranes of the cells are the ribosomes on which proteins are synthesized. The mitochondria contain the enzymes responsible for the final oxidation steps in glucose catabolism and the synthesis of ATP. Some of the small vesicles (called *lysosomes*) are packed with enzymes which, if released into the rest of the cell, would quickly prove lethal, for they can cause many of the macromolecules which make up the cell's structure to degrade into their components. These vesicles function as intracellular scavengers, mopping up unwanted molecules – but they can also act as a sort of cellular suicide pill.

Thus any individual cell has a complex internal set of components. Each of these components represents a separate compartment within which relatively segregated sets of reactions can occur. Communication between these compartments, in the form of exchanging substances and signals, takes place through selective membranes, which act as gatekeepers. This, for instance, is how the 3-carbon acids which are the products of the first stage of glucose breakdown enter mitochondria across their membranes, to be oxidized in a highly ordered sequence of reactions catalysed by enzymes embedded in the internal membranes of the mitochondria. The ATP produced during this oxidation, together with the carbon dioxide which is the final oxidation product, leave the mitochondria again, the ATP to do its business within the cell, the carbon dioxide to be expelled across the external cell membrane. Similarly, signalling molecules and ions such as calcium enter the nucleus through its membrane, carrying information determining which particular sections of DNA are to be transcribed into RNA; the transcribed and edited RNA exits the nucleus, carrying its message in turn to the ribosomes in the cell cytoplasm. Small inorganic ions play a key regulatory and signalling role in these trans-membrane processes.

Homeodynamic order within the cell is thus maintained not merely through the self-stabilizing properties of metabolic webs, but through internal structural constraints set by semipermeable lipid membranes in which are embedded proteins that recognize and regulate the entry and exit of key metabolites. This regulation and recognition is itself modulated by ions such as calcium, and transient modifications to the structure of the proteins themselves (for instance, by transferring the phosphate of ATP onto one of the constituent amino acids of the protein chain).

And indeed, these inorganic constituents of the cell, the calcium, magnesium, sodium, potassium and phosphate, play a crucial role in maintaining the internal environment, which is vital not merely in controlling the activity of enzymes, whose speed of reaction is affected (as discussed earlier in this chapter) by pH and ion concentration, but in general in keeping all the cell's proteins in their three-dimensional, tertiary structure (as described way back in Chapter 2). Changing their immediate microenvironment changes the ways in which protein chains fold and curve around themselves, their shapes in space, and hence also their functions.[17] So the functioning cell, as a unit, constrains the properties of its individual components. The whole has primacy over its parts. This inherent dynamism of the cell is belied by the apparently rigid and fixed structures that are created by the brutalizing techniques of electron microscopy. There are however techniques which enable one to observe in some detail what is going on inside living rather than pickled cells,[18] and the picture which then emerges is as dramatically different as a video record of one's children at play is to a family photo album. Far from being static and immobile, the internal components of the cell are in constant motion. The nuclei spin gently; mitochondria move gracefully through the cytoplasm, occasionally budding off daughters; streams of small particles are in constant migration. All is motion – the traffic and interaction of dynamic order.

How are these internal structures created? Are the motion and composition of each specified down to the last detail by instructions from the genes, or are they selected by the environment, or are they – like the multicellular organisms they compose – the results of autopoiesis? The answer, like the other answers this chapter has given, is that all three processes are at play. Without the genes, of course, the particular amino acid chains that constitute the proteins could not be synthesized. How the chains fold is, as emphasized in the last paragraph, affected by their microenvironments. But this folding has structural constraints, and represents the creation of higher-level orders, given by secondary and tertiary – and even quaternary – structures, than those of the amino acid sequences of which they are composed. The folding patterns and resultant shapes are not simply implicit in or predictable from the sequences: they depend on the environments as well.

Many of the particles visible within the living cell are complexes of numerous proteins wrapped around each other to form giant multi-enzyme assemblies. The most striking of these are the ribosomes. As mentioned in the last chapter, ribosomes contain more than 80 different proteins, along with RNA sequences. The proteins have been isolated, purified and some of them sequenced. But here's the interesting thing. If one takes the individual proteins that constitute the ribosome, and mixes them together in a test-tube in the right environmental conditions, they spontaneously assemble themselves into ribosomes once more. This property of self-assembly is the key to understanding how cells are able to build themselves. It arises as a result of the physical forces acting on the specific proteins of the assembly, driving them to bind together in ways which conform to 'least-energy' configurations (the maths and thermodynamics are complicated and only partially understood, and need not concern us here). Ribosomes are but one example of such self-organizing properties. I have already referred, in Chapter 4, to the way in which actin and myosin, the major muscle proteins, can assemble themselves into contractile filaments. Cells retain their shape by virtue of an internal 'skeleton' composed of fine tubules (*microtubules*) whose principal constituent is the protein tubulin. Microtubules too will spontaneously assemble from a tubulin solution provided the ionic composition is correct (Figure 6.10), and indeed in living cells they can be shown to be undergoing periodic oscillations between their assembled (polymerized) and disassembled forms. Similarly, the ubiquitous lipid and protein membranes, in so many ways vital to both the origin and the preservation of cells, will form spontaneously, like oil films on water, without the need for specific genetic instructions

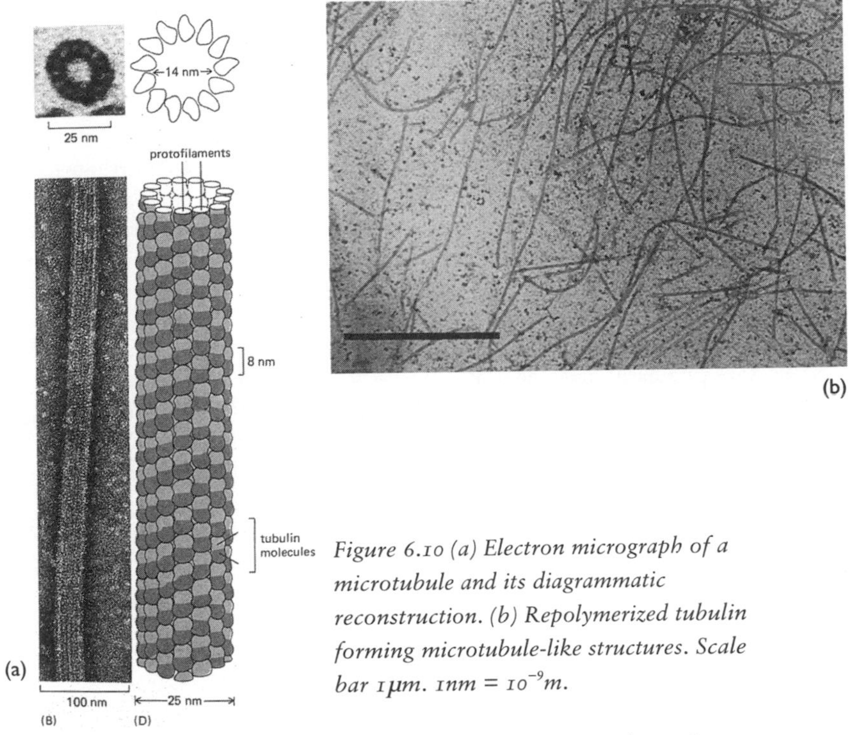

Figure 6.10 (a) Electron micrograph of a microtubule and its diagrammatic reconstruction. (b) Repolymerized tubulin forming microtubule-like structures. Scale bar 1μm. 1nm = 10^{-9}m.

– an intrinsic molecular property which turns out to be at least as important for the origin of life as do the famous replicating molecules of DNA and RNA themselves.

Lifelines, then, are not embedded in genes: their existence implies homeodynamics. Their four dimensions are autopoietically constructed through the interplay of physical forces, the intrinsic chemistry of lipids and proteins, the self-organizing and stabilizing properties of complex metabolic webs, and the specificity of genes which permit the plasticity of ontogeny. The organism is both the weaver and the pattern it weaves, the choreographer and the dance that is danced. That is the fundamental message of this chapter, and therefore in many ways of this entire book. And it provides the framework within which I turn now to consider the mechanisms of evolution.

NOTES

1. For an interesting example of the debate such a proposition arouses, see Brian Goodwin and Richard Dawkins, 'What is an organism?'.
2. C. B. Blakemore and R. C. van Sluyters, 'Reversal of the physiological effects of monocular deprivation in kittens . . .'.
3. Edelman has written a trilogy of books developing this theory and expanding it into a general mechanism which accounts for everything from ontogeny to memory and consciousness. The three, *Neural Darwinism* (1987), *Topobiology* (1988) and *The Remembered Present* (1989), have also been abridged into a more popular work, though it is still tough to read because of his peculiarly convoluted style: *Bright Air, Brilliant Fire* (1992).
4. Francis H. C. Crick, 'Neural Edelmanism'.
5. Rita Levi Montalcini, *In Praise of Imperfection.*
6. Lewis Wolpert provided the general model for this type of pattern-forming development many years ago, with what he called the 'French flag model'. This was subsequently refined by Brian Goodwin (see his 1963 book *Temporal Organisation in Cells*), who pointed out that rather than a continuous gradient, one that pulsed over time provided better three-dimensional control. On the specific issues of axonal growth and patterning discussed here see Dale Purves, *Neural Activity and the Growth of the Brain*, and Josef P. Rauschecker and Peter Marler (eds), *Imprinting and Cortical Plasticity.*
7. Semir Zeki, *A Vision of the Brain.*
8. Purves, *Neural Activity and the Growth of the Brain.*
9. R. L. Smith (ed.), *Sperm Competition and the Evolution of Animal Mating Systems.*
10. Bonnie Spanier, *Im/partial Science.*
11. The term 'constructivist' has a variety of meanings. Closest to mine in this context is that introduced originally by the evolutionary and developmental psychologist Jean Piaget with his concept of genetic epistemology; see e.g. his *Behaviour and Evolution.*
12. I don't know whether 'homeodynamics' is a term I have invented, or whether it has a prior history in biological thinking. Lynn Margulis discusses the same concept but uses instead the term 'homeorrhesis', referring to regulation around a changing set point; see Margulis and Oona West, *Gaia and the Colonisation of Mars.* The neuroendocrinologist Bruce McEwen uses the term 'allostasis' in a similar context.
13. Steven Rose, *The Chemistry of Life.*

14. Henry Kacser and J. A. Burns, 'Molecular democracy: Who shares the controls?'; the passage quoted is from p. 1151.
15. Stuart Kauffman, *At Home in the Universe.*
16. Benno Hess and Alexander Mikhailov, 'Self-organisation in living cells'; Albert Goldbeter, *Biochemical Oscillations and Cellular Rhythms.*
17. Daniel L. Minor Jr and Peter S. Kim, 'Context-dependent secondary structure formation . . .'.
18. Such as, for instance, video-phase contrast microscopy in tissue culture.

7

Universal Darwinism?

Nothing in biology makes sense except in the light of evolution.

Theodosius Dobzhansky

DARWINIAN JUSTIFICATIONS

Some fields of creativity and scholarship live always in the shadow of their own past. It is hard, for instance, for novelists to write or artists to paint, or for the rest of us to read or view their work, without being consciously aware of how previous explorations of the written and visual worlds precede and even overshadow all current work. Natural science is different. It looks forward, not back, and takes casually and for granted the achievements of its ancestors. The shelf-life of a research paper in molecular biology is rarely greater than a couple of years; a 'classic' experiment may be as little as five or even ten years old. Past that age, papers and books become of interest only to historians. Even the names of earlier generations of researchers are forgotten unless they have been eponymized into a piece of equipment (Warburg manometer), a technique (Ringer solution), a mechanism (Krebs cycle) or a unit (volt). Mendel's ratios may be the starting-point for teaching genetics, but they are hardly themselves the focus of current research or debate.

One of the few exceptions to this rule, at least among biologists, is Charles Darwin. He, and the 'ism' to which his name has become attached, crop up so regularly these days that it has even become possible for philosophers to speak of something called 'universal

Darwinism'. The intellectual ferment that surrounds the varying interpretations of Darwinism is as fecund of newspaper articles, polemical tracts and weighty philosophical tomes as it was in the decades after the first appearance of *The Origin of Species* in 1859. The situation could not be more markedly different now from that during the long decades at the beginning of the twentieth century, when Darwinism was in eclipse.

In the years following its first appearance, Darwinism was seen variously as justifying imperialism, racism, capitalism and patriarchy; as symbolizing the death of God and religion; as demystifying humanity; as merely the projection of the social expectations of a Victorian gentleman onto the non-human living world; as providing a universal mechanism for evolution so simple that Darwin's disciple and prophet T. H. Huxley remarked when presented with it 'How stupid not to have thought of that.'

Today, journalists refer to boardroom struggles and takeover battles for companies as 'Darwinian'. Fundamentalists, Christian, Islamic and Jewish, publish learned tracts invested with as many of the trappings of scientificity as they can muster, claiming that evolution cannot account for life on Earth or the human spirit, and attack both Darwin and his followers as doing the work of the Devil. Equally passionate Darwinian protagonists offer a 'tough-minded' ultra-Darwinism as a universal mechanism to explain all phenomena of life. Philosophers follow them; the philosophy department at the London School of Economics offers a popular series of Darwin Seminars, while Daniel Dennett writes a book entitled *Darwin's Dangerous Idea* in which Darwinian mechanisms are described as a 'universal acid' which eats away at everything it touches.[1] Indeed, he proposes that Darwinian mechanisms replicate like viruses, but in all manner of unlikely hosts. Nobel prize-winning immunologist Gerald Edelman interprets the brain processes concerned with experience, memory and consciousness as representing 'neural Darwinism'. Philosopher of science David Hull claims that scientific theories themselves win or lose the struggle for acceptance according to Darwinian mechanisms. One reads of 'Darwinian psychology', 'Darwinian psychiatry', 'Darwinian medicine', 'Darwinian economics'. Richard Dawkins, characteristically, caps the lot with his claim that human culture itself operates on Darwinian principles, the units of transmission being not genes, but 'memes'. Nor are the historians inactive. While preparing this chapter in Goteborg, in Sweden, I was handed a 300-page thesis exclusively concerned with discussing not Darwin himself, but controversies among historians of science as to how to interpret Darwin. Truly, if evolutionary success is to be measured not in the perpetuation of one's genes but in the perpetuation of one's name, Charles D is by current standards a star performer. (He didn't do so badly in the former category either, siring seven children who survived to adulthood and produced an ever-increasing swarm of later descendants – unlike his sterile cousin Francis Galton, whose eugenic dreams failed in personal practice.)

In this and the following chapter I want to look at some of the debates within biology which have surrounded both evolution and natural selection, Darwin's own theory of evolution's mechanism. I shall try to show how, just as with the term 'gene', the simple-minded propositions that these days often go under the name of 'neo-Darwinism', but which I will refer to as *ultra-Darwinism*, are either partial or mistaken. I shall also suggest that it may be time to try to rescue Darwin from some of his over-solicitous modern friends, if we are to do justice – but no more than justice – to the part he and his ideas have played in the history of biology and in our understanding of living processes. To set the issues into context however, it is necessary to begin not with Darwin himself, but with his precursors. I shall then turn to Darwin's own propositions, and the three main problems – of the origins and persistence of variation, of adaptation and of speciation – that he left to his followers to resolve. I leave to the chapter that follows consideration of alternatives to ultra-Darwinism.

THE GREAT CHAIN OF BEING

Before Darwin, the interpretation of life on Earth was trapped within a mode of thinking imposed by biblical traditions. Common observation shows that the living world is divided into different types of animals and plants, and that these differences were maintained across generations. Lions mate and give birth to lion cubs, sheep mate and give birth to lambs. Cubs and lambs in due course grow into lions and sheep and mate in their turn. But lions do not mate with sheep, they eat them; at best, in the paradisiac vision of the Bible, they might lie down peacefully with each other. Even if such similar animal types as horses and donkeys can mate, the result is a sterile cross, in this case a mule. Similarly with plants: nasturtium seeds turn into nasturtiums, hazel nuts into hazel trees. Thus each type or species was believed to be qualitatively distinct and to breed true – a Platonic natural kind, that is. According to biblical myth, life on Earth began during the seven days of Genesis, when God individually created the progenitor pair of each species. These proliferated until the days of Noah's flood, when breeding pairs of all the world's species boarded the Ark and were thus spared to begin the process of repopulating the Earth once the floodwaters had subsided.

The eighteenth century in Europe was the period of the Enlightenment, of the great systematizers and classifiers. The French worked on their vast *Encyclopédie*. Away in Uppsala in Sweden, the botanist Carl von Linné (known as Linnaeus) began the task of classifying all living species. A species was defined as a distinct group of creatures resembling one another in form and capable of fertile mating. Clearly, some species more closely resemble one another than they do other species, so they can be grouped together as, say primates (which include chimpanzees and gorillas) or ungulates (which include sheep and cows). But both primates and ungulates share with many other species the property of giving birth to live young (mammals), and with still more species the property of having a backbone (vertebrates). And so on. Related organisms could be assembled into nested groups, species within genera within families within orders within classes within phyla within, finally, the great kingdoms of animals, plants and fungi (bacteria only got classified later). But all species, however closely interrelated, were regarded as immutable. They had persisted from the beginning and would continue until the end of time. Furthermore, all could be arranged upon some absolute scale of perfection, a Great Chain of Being, beginning with the lowliest and ending with that acme of God's creation, Humankind (Man) himself.

EVOLUTION

Enlightenment stability was not to last, however. Change was in the air, with the quickening pace of the Industrial Revolution. Human intervention, it was clear, could transform the appearance of species, domesticating and producing new varieties of sheep, cattle and dogs, although within a species even the most bizarrely differing varieties – great danes and dachshunds, for instance – are capable of fertile mating, however awkward the mechanics may prove to be in practice. This was also the period of intense interest in geology, not least because of its relevance to the extractive industries of coal and iron. As geologists explored the surface of the Earth and studied the strange objects that miners brought forth from its depths, they began to discover fossils – the petrified remains of mysterious organisms at the same time both like and unlike those currently alive on Earth. Their existence in defined rock strata enabled them to be assigned dates, stretching back many millions of years. Perhaps species were not stable at all. Some living forms which had existed in the past did so no longer. But could they have been ancestors of present forms, into which they had gradually been transformed? This might account for all the family similarities which Linnaean classification had systematized.

Evolution simply means change over time (in fact, it shares a common etymological origin with the term 'development'), and by the beginning of the nineteenth century the arguments that species had indeed evolved – that is, changed over time – and that species currently alive were related, both to fossil ancestors and to one another, were relatively commonplace, at least among the freethinking intelligentsia. Erasmus Darwin, Charles's grandfather, a wealthy country doctor, amateur poet and botanist, argued thus. And so, above all, did the Paris-based naturalist and philosopher Jean-Baptiste Lamarck. Lamarck went further, seeking to offer a mechanism by which evolutionary change might conceivably occur. He found it in terms of individual life experience. Each creature strives to survive, and to do so must endeavour to improve its capacities and skills. Thus, in his famous example, an early ancestor of the giraffe, endowed with only a relatively short neck, could be imagined as stretching up to reach the leaves of the trees on which it fed and thus lengthening its neck, if only imperceptibly. This imperceptible lengthening would then be transmitted to the giraffe's progeny, and over the generations giraffes with ever longer necks would appear.[2]

Lamarck's mechanism has been the butt of cruel jokes by Darwin's advocates for more than a century now, despite periodic attempts by more flexibly minded biologists to revive or even test it. Where it collapses is in the repeated failure to find reproducible evidence that characters acquired during an organism's lifetime can thus be perpetuated, except in certain rather ambiguous and highly constrained test-tube experiments. As the child of fairly orthodox Jewish parents, I was circumcised at birth, just as all other Jewish and Muslim males have been for generations. But the fact that for some four thousand years and two hundred generations my male ancestors had been circumcised did not (as far as I know!) have any effect on the length of my foreskin. Such examples are commonly cited to disprove Lamarck, though since an eight-day-old Jewish boy doesn't exactly strive to have his foreskin removed it isn't exactly what Lamarck had in mind. His model does require some positive effort on the part of the animal. None the less, it is the failure of Lamarckism that lies behind Crick's formulation of his Central Dogma: 'once "information" has passed into the protein it *cannot get out again*'.

NATURAL SELECTION

Charles Darwin did not invent or even demonstrate evolution, although he inferred that it must have occurred.[3] His achievement – and its simultaneous discovery by his contemporary, Alfred Russel Wallace[4] – was to provide a more plausible account of how evolution might take place than that offered half a century previously by Lamarck. Both Darwin and Wallace were above all great observers of the living world, and their observations were enhanced in the course of their travels to lands previously unknown to most Europeans. Wallace earned a living satisfying the wealthy Victorian bourgeoisie's passion for collecting, by trapping tropical birds and butterflies which could be stuffed or preserved and sent home to fill the collectors' mahogany-and-glass cabinets. Darwin's five-year voyage on the survey ship *The Beagle*, as naturalist and companion to the captain, Robert FitzRoy, took him to the rich forests of South America, the Galapagos, and the islands of the South Pacific. He finally settled down near London to spend the next half-lifetime corresponding with plant and animal breeders to learn what they could tell him of the methods and results of their artificial selection procedures.

What Darwin learned from the breeders was that if they mated two animals, the offspring, though similar, were not identical. If one animal in a litter showed a character the breeders were looking for, and they selected that animal to mate with another showing a similar character, there was a chance that the selected character would not only be more common in the next generation, but that with appropriate further selection it could even be enhanced. If nature worked as the artificial breeders did, and over many generations giraffes with longer than average necks mated only with other long-necks, might not nature do for giraffes what breeders did for spaniels and pouter pigeons?

The final clue as to mechanism, at least for Darwin, is supposed to have been provided by his reading of the influential essay on human population by the gloomy Reverend Thomas Malthus, which originally appeared in 1798 and in many editions thereafter, culminating in the sixth in 1826. According to Darwin's notes, he first read this edition and recognized its significance in 1838.[5] Malthus's essay was essentially a moral proposition. The human population, he pointed out, has the capacity to increase in geometrical proportion. Thus if every couple rears four children into adulthood, in the second generation the four become sixteen, in the third the sixteen become sixty-four, and so on. On the other hand, historical records showed that the efforts of human agricultural labour to increase the production of food could do so only in arithmetical proportion (two become four become six become eight . . .). So the availability of food would fall inexorably behind the number of mouths needing to be fed, and there would be a brutish and increasingly desperate struggle for existence.

Malthus saw this as pointing to the inevitable failure of welfare measures, Poor Law relief or charitable attempts to alleviate the lot of the poor, which would merely encourage their intemperate breeding practices (the current backlash against 'welfare mums' in both Britain and the USA has distinct Malthusian undertones). But for Darwin it provided the missing line in his syllogism, the core of his famous theory:

1. Like breeds like, with variations.
2. Some of these varieties are more favourable (to the breeder, or to nature) than others.
3. All creatures produce more offspring than can survive to breed in their turn.
4. The more favoured varieties are more likely to survive long enough to breed.
5. Hence there will be more of the favoured variety in the next generation.
6. Thus species will tend to evolve over time.

This process is natural selection as Darwin described it. As a syllogism it has a compelling logicality: if 1, 2 and 3 are true, then 4, 5 and 6 follow inevitably. This is why philosophers such as Dennett are able to describe natural selection as a universal mechanism, applicable whether one is talking about living organisms or computer viruses. Thus formulated, it constitutes one of the few specifically biological laws, to be ranked alongside the great universals of physics. And this is why Huxley kicked himself at his stupidity at not having seen it for himself.

That Darwin's ideas did not win immediate and universal assent among biologists or geologists was only in part because of the threat they presented to orthodox, mainly Christian views in the shape of the suggestion that humans might be related, however distantly, to other living primates.[6] Despite the care with which Darwin marshalled his arguments, and the wealth of observational data he presented, there were major theoretical difficulties at the heart of the theory. Although, as a syllogism, natural selection might be unassailable, three central problems remained. The first was the mechanism of transmission of both similarities and variations. The second was the classic argument from design: how could gradual change result in such seemingly perfectly adapted structures as the eye ('What use is half an eye,' the critics asked). The third was the problem of speciation. Today, the first is no longer a problem, the second raises a number of important conceptual issues, and the third is still with us. At the same time, as will become apparent later in this chapter and in the next, new debates over the meanings and significance of Darwinian mechanisms are constantly emerging.

WHAT DARWINISM DID

Before looking at the problems, let us be clear about the achievements of evolutionary theory and the mechanism that Darwin proposed, for even today these are persistently misunderstood. First, it demolished for all time the idea of the immutability of species, and, even more importantly, of a Great Chain of Being. Humans are no longer at the pinnacle of Creation. Instead, as in Darwin's own metaphor, the relationships between living forms can be depicted as the branches and twigs of a tree (Figure 7.1). Humans are at the end of one twig, and all other current living forms are at the ends of others. Some, like chimpanzees and gorillas, are a mere one or two twigs away from us. Others, like slugs, wasps, mushrooms and amoebae, may be separated by many branches. But there is no way in which any currently living form can be described as 'more' or 'less' evolved than any other. It is worth pointing out that in his book *Wonderful Life*, Stephen Jay Gould has criticized the conventional iconography embodied in tree-like representations like Figure 7.1, on the grounds that it implies an evolutionary world of increasing diversity over time, as if the early Earth was populated by only a very few living forms. It may well be that only a fraction of the living forms at any one time on Earth will be the ancestors of others – lines, species, phyla may die out without leaving progeny – but this does not necessarily mean an increase in diversity in the way the tree-diagrams imply.

All of us currently alive, amoebae as well as humans, are thus equal in the sense that we are all the current products and successful survivors of evolutionary history. Thus, despite the common parlance, itself left over from pre-evolutionary days, there is no scale of life on the basis of which one can judge some currently living forms as 'lower' and others as 'higher', more or less 'evolutionarily successful'. The fact that we and oak trees and cholera bacilli are all here together means that we are all survivors: there is no judging between us, no taxonomic order of merit which ranks us humans above the rest. This is so difficult a thought that not even all biologists grasp it; I cringe every time I read a biology textbook which casually throws around terms such as 'lower' and 'higher' in 'the evolutionary scale'.

There is an alternative higher/lower scale which is sometimes employed – that of complexity. While we may all be equally evolved, surely humans are more complex as individual living forms, and have developed more complex forms of society, than bacilli. Intuitively this seems true, although the definition of complexity in this sense is not easy. An interesting attempt has been made by the developmental biologist John Tyler Bonner, who suggests describing the relative complexity of living organisms in terms of the number of different specific cell types they contain. On this score we humans, with more than 250 identifiably different cell types in our bodies, rank much higher than the segmented worms in whose body plan the same limited number of cell types repeat over and over, and higher than giant trees many times larger in mass than ourselves, but more simply constructed at the cellular level.[7] It has even been argued that evolution necessarily proceeds in the direction of greater complexity,[8] but there are many life forms which seem to do very well with rather simple body plans and small numbers of different cells. Neither complexity nor brains can be said to be inevitable products of evolutionary trajectories, but, as I argued in Chapter 3, once a species has taken even a small, tentative step along the path to a nervous system and a brain, it will find itself under considerable evolutionary pressure to continue along that path.

PEDIGREE OF MAN.

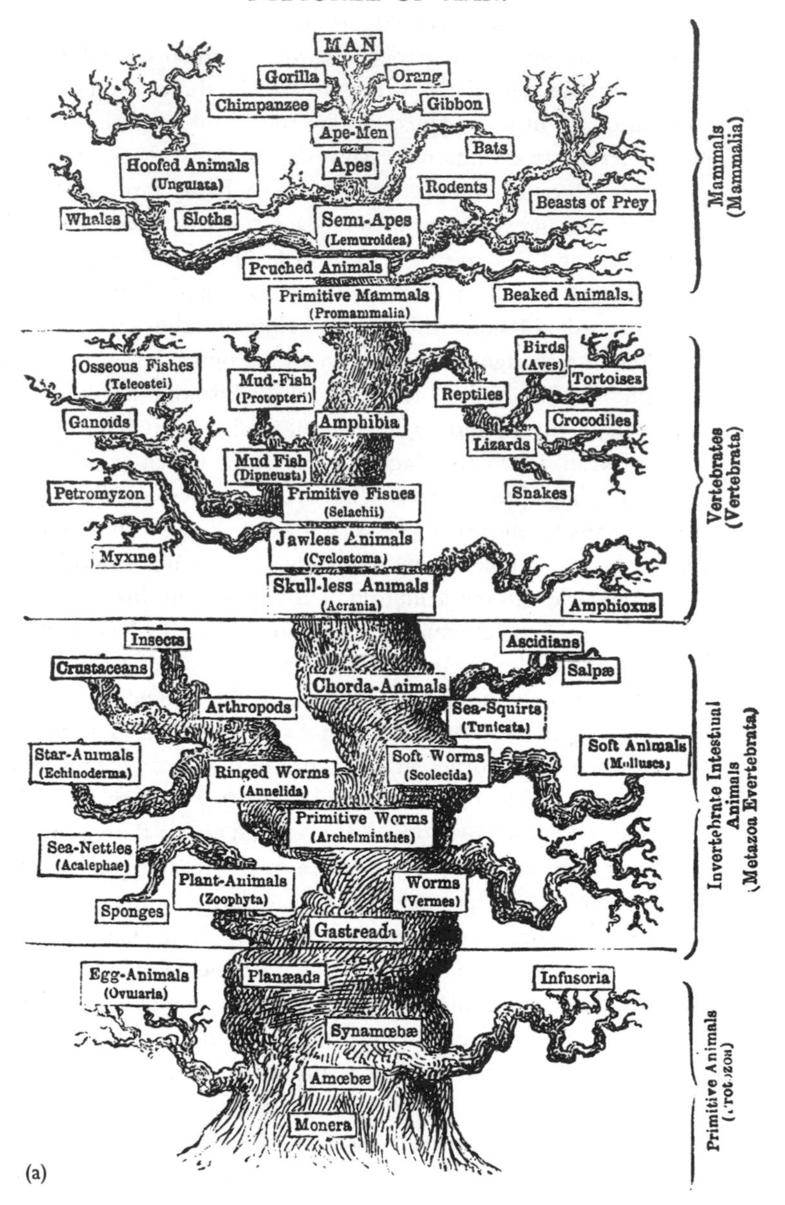

(a)

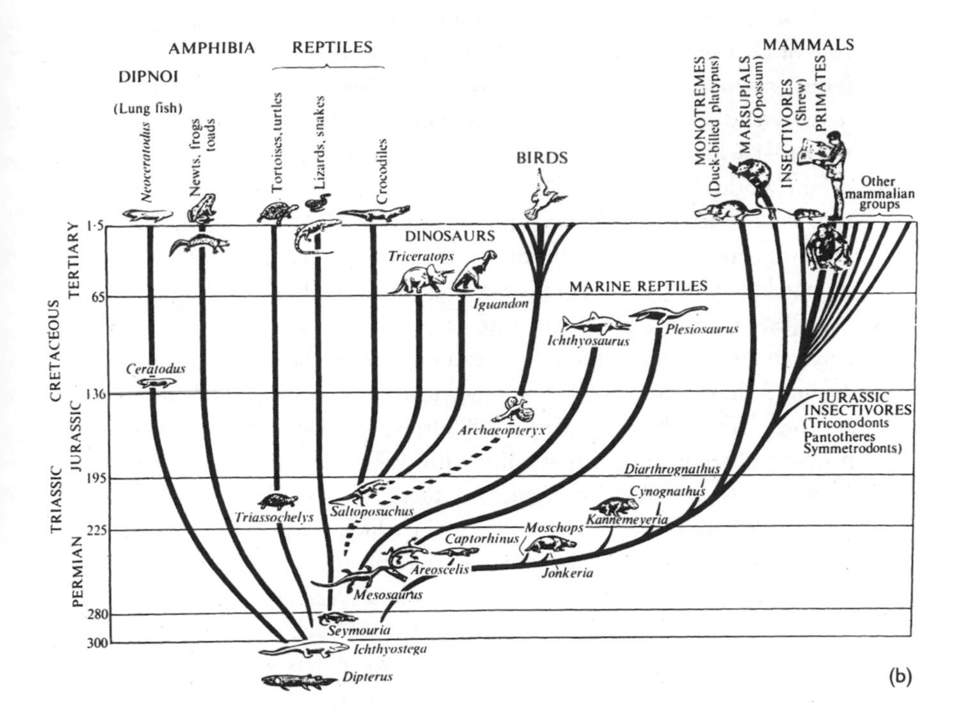

Figure 7.1 Evolutionary trees, ancient and modern:
(a) Ernst Haeckel's 'Pedigree of Man', 1879;
(b) as depicted by J. Z. Young in 1971.

A further crucial feature of Darwinism is that it acknowledged the role of chance in a way that earlier scientific theories had not. This is one reason why it was anathema to some of Darwin's contemporaries in the scientific community of Victorian England, brought up to respect the order which physics and chemistry seemed to be able to impose on the world. 'The theory of higgledy-piggledy' was how that grand old man of Victorian science, John Herschel, put it.[9] Natural selection abolishes purpose from evolution, and in consequence, some felt, from human life itself. Later generations of religiously inclined evolutionary biologists therefore sought to restore purpose and direction to the evolutionary process – the classic example being the Catholic Pierre Teilhard de Chardin, with his 'omega point' towards which life is striving.[10] The ethologist William Thorpe precisely summed up the problem of reconciling Christianity with evolution when he called his book *Purpose in a World of Chance*.[11] Others, of course, have embraced the austere grandeur of a world-view in which purpose is imposed by humanity, not read off from nature – no one more so perhaps than Jacques Monod in his book *Chance and Necessity*.[12] I shall not dwell on this point further here, but shall come back to it in the next chapter by way of Popper's contrast between passive and active Darwinism. Let me for now return to the problems which Darwin's theory bequeathed to his successors.

THE ORIGINS AND PRESERVATION OF VARIATION

Although the first problem, that of the mechanism of transmission, greatly vexed Darwin's contemporaries and followers, it now has a fully satisfactory solution. It arose at the time because there was no concept of the gene. Breeders could ensure favourable outcomes by combing populations for variants they preferred and then controlling their mating. But even if the occasional favourable variation emerged by chance, in nature it would be most unlikely to find a mate which shared the same favourable variation. And if characters blended during mating, as Galton's studies of continuous variation suggested they did, then unusual variations, however desirable, would rapidly be diluted out.

Darwin insisted on gradualism: for him variations were minute, and change did not occur by large, sudden leaps. Many of his supporters pressed him to accept such leaps (major mutations, as they would now be regarded) as the only way to save his theory, but he declined. Changes had to be gradual: it was after all evolution, not revolution. Indeed, in the absence of any alternatives he had even begun to contemplate Lamarckian mechanisms by the time, towards the end of his life, he came to prepare the final editions of *The Origin*. The problem of the preservation of favoured characteristics remained unresolved until the rediscovery of Mendel's work two decades after Darwin's death, and by that time natural selection theory had begun to fall into disrepute, precisely because it could not resolve these difficulties. As I mentioned in Chapter 5, the early decades of the twentieth century were those when Mendelism triumphed, and it wasn't until 1930 that an acceptable synthesis of Darwinism and Mendelism became available. If small variations were the result of changes in Mendelian genes, then they would be preserved, and if dominant they would recur in subsequent generations. Even if recessive, they would not disappear, but lie latent until two individuals with the same recessive genes mated, in which case they would be expressed phenotypically in a proportion of the offspring. Thus, properly understood, Mendelism and mutation provided the mechanism for the preservation and perpetuation of favourable change that Darwinism required. Neo-Darwinism, or the modern synthetic theory as it became known, was given mathematical expression in books appearing more or less simultaneously by the statistician Ronald Fisher and the polymath physiologist, biochemist and geneticist J. B. S. Haldane in Britain, and Sewall Wright in the USA. As will become apparent in the next chapter, however, there was a crucial difference in the way in which Fisher and Haldane on the one hand and Wright on the other approached the synthesis, a difference whose consequences lie at the heart of current disputes over the basis on which selection can occur.

A DETOUR THROUGH HERITABILITY

Fisher worked at a plant research institute, Rothamsted, and behind his synthesis of Darwinism and Mendelism lay the need to understand the nature and origins of variation in populations. Plant a field with a genetically homogeneous variety of wheat, and treat patches of the field with varying combinations of fertilizers, soil quality, availability of water, and so on, and crop yield will also vary. How much of that variation is due to genetic differences, and how much to the different environments? In an absolutely uniform environment, of course – were such a thing possible – all the variance would be contributed by the genes, and with absolutely identical genes all the variance would be contributed by the environment. But this never happens. Genotypes and environments both vary, and the purpose of heritability estimates (see below) is to try to tease them apart. While, as should be apparent from the arguments of the previous chapters, one cannot ask how much of the growth of any single plant is the result of genes and how much of environment, it is possible to ask a similar question about differences between individuals in populations. To do so, however, it is necessary to make some simplifying assumptions. To begin with, variance is given a rather more rigorous statistical definition, in order to describe the way in which any particular measure of a trait in a population is distributed about the mean value for that population. It is assumed to be made up of a component contributed by the genes and a component contributed by the environment, which can simply be added together to give a total of nearly 100 per cent. The remainder, which to make the mathematics work has to be a rather small proportion of the total, is considered to be the product of an interaction between genes and environment. To put it in the form of an equation, if V is the total variance, G the genetic contribution and E the environmental contribution, then:

$$V = G + E + (G \times E)$$

If genotypes are distributed randomly across environments, it is then possible to estimate the value of a quantity called *heritability*, which defines the proportion of the variance which is genetically determined. A heritability of 1.0, or 100 per cent, indicates that in this particular environment all the variance is genetic; 0.0 indicates that it is all environmental. However, the mathematics works only if all the relevant simplifying assumptions are made. If there is a great deal of interaction between genes and environment – if genes behave according to Dobzhansky's vision of norms of reaction, if genes interact with each other, and if the relationships are not linear and additive but interactive – then the entire mathematical apparatus of heritability estimates falls apart. As J. B. S. Haldane pointed out back in 1946, in general '*m* genotypes in *n* environments generate $(mn)!/m!n!$ kinds of interaction'.[13] For the non-mathematical, consider simply 3 genotypes and 3 environments. Then mn is 9, and $(mn)!$ (which means $9 \times 8 \times 7 \times 6 \times \ldots$) is 362,880; $m!$ and $n!$ are each $3 \times 2 \times 1$, or 6, and the number of interactions is no fewer than 10,080.

From everything that I have been arguing in the last two chapters, it will be seen that, like Mendel's laws themselves, the meaningful application of heritability estimates is possible only in very special cases, and the majority of traits of interest outside the special world of artificial selection are unlikely to number among them. Furthermore, without going into the technical details of the mathematics, the figure derived for the heritability is itself dependent on the environment – that is, if you change the environment, the heritability estimate changes.

These caveats perhaps help to explain why, more than any other aspect of genetics, heritability estimates have been so persistently misunderstood, often by other biologists and especially by those psychologists whose goal is to provide precise measures of human attributes (psychometricians), to say nothing of the non-specialist public.[14] The estimate works only if the simplifying assumptions are valid; the figure obtained applies not to an individual but to differences within a randomly interbreeding population, and cannot be applied to differences between populations;[15] it assumes the distribution of genotypes across environments to be random, and the estimate changes if these environments are changed.

So why bother with it at all? The answer is that if you are a plant or animal breeder and want to know about crop yield, or milk yield in cows, it can provide valuable information. Where it becomes wholly misleading is when efforts are made to apply the same sort of estimates to aspects of human behaviour. Milk yield is a phenotype which is reasonably straightforward to measure. But intelligence? Political tendency? Likelihood of getting divorced? Religiosity? Job satisfaction? Impulsiveness? Ease of making friends? Taste in clothes? And if such phenotypes are problematic, just what is meant by 'the environment' in such equations? As I have argued, a gene's environment can be understood at many levels, from those of the rest of the genome, to the cell, to the developing organism, to the natural and, for humans, social world within which that organism is embedded. None of this matters to those who insist on applying the heritability equations; for them 'environment' is simply an undefined portmanteau term, as abstracted from living reality as are the genes to which it is counterposed.

Ever since Fisher, psychometricians and human behaviour geneticists have attempted to apply heritability statistics to human attributes such as those mentioned above. As one cannot treat human populations in quite the same way as when conducting breeding experiments with wheat or cattle, and distribute genotypes across environments, one has to make do with what nature and society provide between them. The standard approach has been to compare traits in siblings and other family members, who have some genes in common, and above all to compare identical (monozygotic, MZ) with non-identical (dizygotic, DZ) twins. MZ twins have essentially identical genotypes, DZs are no more genetically alike than any same-sex sibling pair. This provides the genotypic distinctions one needs. How about the environmental variation? The trouble is of course that most siblings share a similar family environment, so similarities detected between them are inextricably the result of both genes and environment.

The 'ideal' experimental situation is the relatively rare one in which identical twins are separated at birth and reared apart, a situation easier to achieve among laboratory rats than among humans. The next best is an adoption study in which one can compare some character in an adopted child with that in his or her adoptive and real parents. The controversies surrounding such studies go far beyond the question of whether the most famous of them, those published by Cyril Burt from the 1930s to the 1950s, were fraudulent (the general consensus, despite a powerful revisionist attempt to rehabilitate Burt in the 1980s, is that, to put it politely, Burt's data cannot be relied upon[16]). The problems are manifold. To mention just two, separated twins tend to be placed in rather similar environments, and are often not really separated at all; while by contrast with the naïve assumptions of the psychometricians, adoptive parents are unlikely to treat their adoptive child 'exactly' as they would a natural one, and are far more likely to be anxiously on the look-out for tendencies which reveal the child to be 'taking after' some undesirable character of its natural parent. Such real-life problems are simply swept aside in the process of fitting the numbers obtained into the complex statistical manipulations required to generate the seemingly objective heritability estimate.

Currently the most comprehensive studies of twins are those based on the register compiled by Thomas Bouchard and his colleagues at Minneapolis–St Paul (a suitable site, given that these are known as the Twin Cities, and their inhabitants are intensely proud of their baseball team, known as the Twins).[17] It is from such studies that relatively high heritability estimates (above 35 per cent) have been derived for such diverse attributes as attitudes to the death penalty, Sabbath observance, working mothers, military drill, white superiority, cousin marriage, royalty, conventional clothes, apartheid, disarmament, censorship, 'white lies', jazz and divorce.[18] Even nudist camps and women judges come in at around 25 per cent, so it is I suppose a matter of some surprise that there appears to be virtually zero heritability for 'pyjama parties', straitjackets and coeducation.

The most parsimonious explanation for this bizarre set of statistics is that they demonstrate the inappropriateness of attempting to apply a mathematical formula devised for plant and animal breeding to such dubious phenotypic characters as the diversity of human social behaviour and attitudes. And, as I have emphasized, even with the phenotypic measures to which they can properly be applied, the estimates are by definition within-population measures. In principle it is possible for all the variance within each of two populations to be genetic, and the differences between the two populations entirely environmental. This possibility is accepted even by those geneticists who give more credence to heritability estimates than I do. Strictly, there is no known way of estimating the heritability of differences between populations. Those who attempt to use the estimates in this way are traducing science in the interests of what is a more or less covert racist agenda.[19] The fact is that until such time as humans live in a society in which social barriers restricting relationships between individuals from different ethnic and social groups no longer exist, such estimates are scientifically meaningless, though they remain socially and politically pernicious.

Yet some psychometricians and behaviour geneticists argue that even such high heritabilities underestimate the true influence of the genes. In his version of the Dawkins 'extended phenotype' argument, Bouchard proposes that our genes 'predispose' us to seek environments congenial to the genetic imperatives.[20] Thus genes create environments, and 'environment' – whatever that term may mean – ceases to be a truly independent variable in the heritability equations. Genes, therefore, are a major cause of everything from childhood accidents to divorce in mid-life, both supposed to be 50 per cent heritable, for such genes lead their owners to place themselves in situations in which the probability of accident or divorce increases. Replace the gene's-eye view of the world with the lifeline perspective which I have been emphasizing, and this insistence on organisms making their own history echoes my own argument. But, like the claims for the 'extended phenotype', it does so by perversely and mistakenly swallowing the four-dimensional universe of lifelines entirely into the double helix of DNA.

Heritability estimates therefore remain a tribute to the enduring power of reductionist thinking within some areas of population genetics, as much as to the political climate which fosters them.[21] Such estimates, however biologically and sociologically impoverished the framework within which they are calculated, and however inappropriately they are applied, even within their own limited terms, are given an apparently scientific gloss because they can be expressed in mathematical, and hence seemingly unchallengeable form. Deference to maths, or rather deference to numerology, strikes again. The real biological issues lie beyond their reach, and I shall not return to heritability again in this book.

ADAPTATION AND DESIGN

For Darwin and his contemporaries, the question of adaptation was even more problematic than mode of transmission, as indeed it still is for some present-day fundamentalist religious critics of evolutionary theory. The problem lies in an argument that precedes Darwin, and is often posed in the form it was originally given by the theologian William Paley in his book *Natural Theology*, written at the very beginning of the nineteenth century. If, on a walk through the countryside, you stumble across a watch lying on the ground, only the briefest examination is needed to convince you that it cannot have come about by chance. The watch and its inner mechanisms show clear evidence of design, and how can one have design without a designer? If this is so for the relatively crude mechanism of a watch, how much more for such marvellous structures as the eye. Darwin himself confessed his terror when trying to think about the possible evolution of the eye. On closer inspection, though, this apparent problem vanishes.

Dawkins confronts this question head-on in *The Blind Watchmaker* and its successors: 'What use is half an eye?' he asks, and answers, 'One per cent better than 49 per cent of an eye, and the difference is significant.'[22] The trouble with this argument is that there is no way of determining whether, among our evolutionary ancestors, 50 per cent of an eye ever proved significantly better in Darwinian terms – that is, whether it contributed significantly more to reproductive success – than 49 per cent. It would depend on what other costs the organism accrued in achieving this 1 per cent advantage, and on how much having eyes contributed to its success in finding food, and avoiding predators so as to increase its chances of finding a mate and hence reproducing. Of course, no such evidence can be forthcoming, and so the claim must remain an undemonstrable assertion, although one which most biologists will find reasonably convincing. Dawkins goes on to cite evidence that serviceable image-forming eyes have evolved independently at least forty times in different invertebrate groups, quite apart from 'the' eye, by which we mean the light- and image-detecting mechanisms which humans share with our more immediate evolutionary neighbours.[23]

How many generations would it take to evolve such an eye from an initial flat retina, above a flat pigment layer surmounted by a protective transparent layer? Dawkins cites a computer model by Dan Nilsson and Susanne Pelger which could do it in under half a million. At the rate of one generation a year, this means just 500,000 years, easily attainable within the timespan of life on Earth. The required assumptions are that each step is heritable, digital in effect, and provides a selective advantage to the creature which carries the variation.

Accepting these assumptions unquestioningly requires something of an act of faith (and, as I argue below, there are grounds for rather less credulity than Dawkins affects to display), but even so I see no problem with the general principle invoked here. In the classical Popperian sense, as we have seen, such evolutionary stories are unfalsifiable. All that we can do, all that we are required to do, is offer plausible accounts of how a process may have occurred or a structure may have evolved, in response to those who claim that it is impossible on *a priori* grounds. If I argue, as I do, that life is a good deal more complex than the computer-generated biomorphs that Dawkins has created as a spin-off from his writings, this should not be read as yielding any ground at all to those who would argue that life is a product of anything other than material forces operating in a material universe, potentially explicable by the methods of a (non-reductive) science.

Once again, as with genetic transmission, the problem of adaptation – at least as it confronted Darwin – is not an insoluble one. He surmised that it could be resolved, given enough evolutionary space and time, and he was surely right. Later on, I shall turn to how the problem recurs in its modern form.

THE LIMITS TO NATURAL SELECTION

The third major problem that Darwin faced, and which his theory in its simple form was unable to resolve, is that of speciation. It may seem extraordinary, but the Darwinian syllogism of natural selection presented on page 181 provides no mechanism for the formation of new species, which was after all ostensibly what *The Origin* was all about. All it says is that, in any given circumstances, external conditions (the environment, nature) will favour the perpetuation of varieties which can do their species-thing a bit better than the rest. Antelopes, for example, are preyed upon by lions. Any antelope in a group which has been evolutionarily favoured by being able to run slightly faster than the others has a slightly better chance of avoiding lions, and is therefore that much more likely to survive. Similarly, lions which can run faster, or develop cooperative methods of stalking their prey in packs, will boost their chances of survival. But this won't in itself turn antelopes or lions, or their descendants, into new species.

There is a real-life example of this type of process in action. It is found in all the textbooks, if only because it is one of the best-documented examples of a change in the form of a species over time which can be attributed to natural selection (as opposed to some test-tube experiments with bacterial populations). The peppered moth, widely distributed throughout Britain, spends much of its time clinging to tree trunks. As its name implies, the normal form of this species is a speckled brown, but a somewhat rarer, black (melanic) variety also occurs, first observed in Manchester in the middle of the nineteenth century. The British enthusiasm for nature study meant that the moth has been observed over many years, and records kept of the proportions of the two forms, which showed a steady increase in the proportion of the dark over the light form in industrial areas in the twentieth century. The moths are much preyed upon by birds, and an obvious interpretation is that in the absence of environmental pollution which darkens the tree bark, the light, speckled form is harder for birds to spot. Where pollution darkens the tree bark, the speckled form will stand out, while the dark form will be better concealed.

In 1955 H. B. D. Kettlewell checked this hypothesis, and showed that the dark form of the moth was indeed at a selective advantage (being less preyed upon) in soot-blackened areas.[24] As expected, the reverse is the case in unpolluted areas. Admittedly the example is not really of natural selection by competition for scarce resources, the original Darwinian motor, but we can allow it none the less. What makes it particularly instructive is that as the shift to less polluting energy sources reduces the amount of soot in the air around Manchester, and trees suffer less blackening, so the melanic form of the moth is decreasing and the speckled form is increasing. Whereas a few years ago blacks outnumbered speckleds by more than 2 to 1, the proportions are now reversed, and the dark form's days seem numbered.

So natural selection *can* work to change populations, increasing the adaptiveness of individuals within them; favoured varieties are preserved and therefore their distribution in the population changes with time. Furthermore, the example of the peppered moth demonstrates another fundamental point about natural selection. By definition, a 'more favoured variety' is one which is favoured *under current circumstances*. Evolution by natural selection can respond only to the current situation – it cannot predict the future. At one point of the species' trajectory in time, it is the speckled form which has the greater survival value, then the melanic, and at a later time the speckled form again. The environmental change occurs, and natural selection trails along behind, following, responding, but never leading – and never predicting.

This inability to predict future advantage, and therefore to adapt in advance, holds even in the lifetime of an individual. A mutation which resulted in the adult antelope being able to run faster, but which also meant that it took longer to mature and was therefore more vulnerable to attack by lions for longer periods, would scarcely have much chance of spreading in the antelope population.

SEXUAL SELECTION

There are two further important twists in the tale/tail of the evolutionary adaptation story. The first concerns sex. If all adaptation serves the function of enhancing survival, how come so many animals – especially males – have traits which seem on the face of it to be inimical to a long and efficient life? The peacock's train is the classic case. How and why does there evolve such an apparently dysfunctional object, of such startling beauty to human eyes? The question vexed Darwin so much that he was led to develop an entire supplementary theory of selection – sexual selection. To pass on their genes, males and females need to mate, and in such animal species that have been studied, given conditions where choice is possible (which means outside the standard laboratory cage), mating is non-random. Potential mates compete in various ways with members of their own sex, and choose a partner of the opposite sex from among a range of potential candidates. What determines success in these two ventures?

Darwin's view was that, by and large, it is the female of the species that does the choosing. He went so far as to postulate that animals have an aesthetic sense, and tend to choose the most beautiful of the potential mates. If peahens regarded the peacock's tail as humans did, they would tend to choose the mate with the most striking tail. Even if one discounts the possibility of aesthetic judgement (or at least an aesthetic judgement which coincides with that of humans, for the sexual adornments carried by the males of many species often strike human observers as more absurd or extraordinary than beautiful), one has only to make the assumption that at some past time, for whatever reason, peahens were attracted to peacocks with bright fan-like tails; this trait would then be selected, and would spread in the male population, and tails would evolve into more and more splendid objects as a result.

According to a slightly different version of sexual selection theory, to grow an elaborate tail requires a considerable expenditure of energy, and, because the tail is manifestly a handicap to normal survival – it makes the bird more conspicuous to predators, and less able to move fast to escape – then any male which survives to adulthood bearing such a burden must be particularly fit in other ways. In this picture the tail becomes a sort of marker, indicating that its possessor is a genetically good bet for a potential partner.

There has been no lack of those who have sought to take the theory, in whichever version, and press it into service to provide an evolutionary 'Darwinian' explanation for human sexual preferences. The general procedure, in this as in so much of the reductive approach offered by the new genetics and sociobiology, is to treat metaphor as if it were homology. For example, competition for mates among human males is discussed as the macro-version of what is said to be the micro-level competition among individual sperms to be 'the one' to successfully penetrate and fertilize the egg. Males and their sperm compete, females and their ova quiescently await their fate.

The problem is that, as with most human extensions of evolutionary mechanisms, but in an even more extreme form, such accounts simply cannot encompass the rich diversity of human experience. Instead they fall back on traditional and often sexist caricatures so crude as to make cheap romantic novels read like sociological essays. Thus the sociobiologists largely ignore the historical and anthropological evidence of variation in social practices across time and space,[25] and instead treat current Western norms (or rather, assertive restatements of what they perceive to be those norms, for they show as little respect for sociology as they do for history or anthropology) as if they were human universals. For example, there have been widely publicized claims that there are universal human standards of beauty. These are based on a cross-cultural comparison of ratings given by Japanese and Western males to computer-generated faces.[26] That the two civilizations have been approaching one another culturally for several generations, and share visual images transmitted via cinema, television and advertising, is not allowed to stand in the way of this drive to evolutionary universalism. Symmetry of feature is apparently highly regarded, and we have even been regaled with tales, based on evidence which would be laughed out of court did it not have the fascination of prurience, that women have more orgasms during sex with men whose bodies are symmetrical.[27] It has to be said that the relevance of this observation to the question of whether there are, as a result of these joyous matings, more offspring – which is after all the only relevant Darwinian criterion – is not specified. On the other hand, adulterous matings are said to have a greater chance of resulting in pregnancy than those within marriage.[28]

As for sexual display, are not the Porsche and the Rolex, still overwhelmingly the appurtenances of financially successful males, the equivalent of the peacock's tail, demonstrating the genetic fitness of their owners to admiring females? The trouble is that wealth is no measure of genetic fitness, and although it may be inherited, the mode of transmission is not via the genes, nor is there much evidence that its possession results in a greater number of offspring. Once again, the supposed Darwinian imperative is negated at its most fundamental level. Sexual selection may be – probably is – an important mechanism by which to account for otherwise improbable features varying from the anatomical, like the turkey's wattle, to the behavioural, such as the bower bird's courtship practices, but we should not let its enthusiasts blind us to the more obvious explanations for the complexity of human sexual arrangements.

ALTRUISM

We now come to the claims for the genetic mechanism and evolutionary significance of altruistic behaviour, and here we are at the heart of sociobiological thinking. The problem for evolutionists is straightforwardly stated. Students of behaviour have described many examples of animals acting in ways which appear not to be in what may be interpreted as their genetic interest. That is, if we assume that organisms seek to maximize their reproductive success, and to pass on as many of their genes as possible to a succeeding generation, then how do we account for birds which, on detecting a predator, draw attention to it and simultaneously to themselves by uttering warning cries to alert the remainder of the flock? Ought they not instead try to make themselves as inconspicuous as possible, so as to diminish the chance of being picked off?

Back in the 1960s, V. C. Wynne-Edwards attempted to account for a different example of seemingly altruistic behaviour. How are the numbers of a particular population of animals regulated, when their territory and food supplies are limited? One argument would be that they all breed to their maximum capacity, and that only the 'fittest' survive the subsequent struggle to obtain adequate food. Wynne Edwards offered an alternative explanation, based on studies of, among other species, grouse. He suggested that grouse have evolved a special display behaviour which informs them of the size of the population, and that individuals then respond to the problem of overpopulation by a sort of self-denial, limiting the numbers of their own offspring for the good of the community as a whole. He called this type of behaviour *group selection*.[29] Evolutionary biologists were quick to point out what they saw as the flaws in his proposed mechanism. Selection, they argued, could act only at the level of the individual, and, for any individual, maximizing the number of its own offspring is the Darwinian driving force. So if most of the grouse were limiting the numbers of offspring they produced, then selection would favour any variant which 'cheated' by trading on the virtuous self-sacrifice of the remainder. So the number of 'cheats' would soon spread through the population, while the numbers who deliberately restrained themselves would fall. Group selection on this basis was a non-starter, although Wynne-Edwards continued to argue his case, against the prevailing climate of opinion.

So how could seemingly altruistic behaviour evolve? The clue is supposed to have been provided in an offhand remark by J. B. S. Haldane, who pointed out that on the basis of the proportion of genes he shared with his closer relatives, he ought to be prepared to sacrifice himself for two brothers, or eight cousins. That is, on the assumption that the life process is all about passing on one's genes to the next generation, then there is a genetic rationality about the individual risking its own life if by so doing it can ensure the survival, and presumably the reproductive success, of a sufficient number of those who share a proportion of its genes. This was a typically bravura statement by Haldane, who throughout his life was very proud of the fact that he had conducted many of his more hazardous physiological experiments – from ingesting excessive quantities of ammonium chloride to measure the effect of changing the acidity of the blood, to testing survival times in the restricted atmosphere of submarines – using himself as a human guinea pig.

However, Haldane's offhand remark was given serious mathematical form by William Hamilton in 1964,[30] and termed *kin selection*. It was E. O. Wilson who, in 1975, brought the argument to the attention not merely of mainstream biologists but of a much wider public as well, when, in a deliberate evocation of the Darwin–Mendel 'modern synthesis' of the 1930s, he called his book *Sociobiology: The New Synthesis*.[31] However, the term that was to take hold in the popular imagination was due not to Wilson but to Dawkins, when the following year he published his evangelizing version of ultra-Darwinian and sociobiological theory: *The Selfish Gene*.[32] (It is worth making clear yet again that Dawkins' genes aren't selfish in the sense in which we might refer to 'gay' or 'aggression' genes. Dawkins' genes do not necessarily confer selfishness on their possessor; they are intended to ensure that their possessor does what is necessary in order that his or her genes are able to replicate and copies can be passed on to the next generation. This may of course include contributing to cooperative behaviour.)

Kin selection, like sexual selection, is a model, a mathematical formulation which, if one grants its basic assumption – that living forms exist primarily to perpetuate their genes – is as inevitable a syllogism as the original Darwinian formulation of natural selection. Although I see no reason to doubt the principle, proving that it applies in any specific real-life case is harder. Certainly, behaviour which might be defined as altruistic does occur among animals living in groups, although equally there is no shortage of evidence that such animals compete with one another. The empirical question is whether apparently altruistic behaviour can be shown to benefit preferentially the kin of the altruist rather than the group as a whole. Considerable evidence in support of this claim has been collected since 1975, but the problem is that in most cases it is open to other interpretations too, despite the prior commitment of kin selectionist theorists to shoehorn the data into their existing theoretical framework. Perhaps in response to the relative lack of experimental support for the kin selection argument, despite its theoretically compelling nature, Robert Trivers drew a distinction between two forms of altruism. One is response to a perceived genetic advantage, and the other is what he called *reciprocal altruism*[33] – an altruistic act performed to benefit non-kin, but in expectation of a subsequent return of the compliment – you scratch my back, I'll scratch yours (literally so with the mutual grooming behaviour of many monkey species).

As with sexual selection, there was no shortage of those, calling themselves human sociobiologists, who maintained that the arguments put forward to account for seemingly altruistic behaviour in non-human animals could be applied to our own condition. You might jump into a river to save a drowning man, even though he was not related to you, on the assumption that if you subsequently got into difficulties while swimming, he might rescue you. As unlikely as this scenario may be, it is one which popularizers of sociobiology have used to describe how reciprocal altruism might work.[34] Once again, a metaphorical relationship has been given the status of homology.

So, is there anything from the annals of human behaviour which might provide an example fitting the sociobiological bill? As with so much else, the quality of the research claims are too impoverished to take wholly seriously. How, for example, might one show that the more genes parents shared with their children, the more care – 'investment' – they would put into them (that is, the more altruistic, in the kin selection sense of the term, they should be)? Parents can share more than half of their genes with their children if the two parents themselves have a proportion of their genes in common. This is a phenomenon called assortative mating. Here is how the argument goes, again in syllogistic form:

1. There is evidence for the heritability of political views.
2. Therefore a couple who both vote the same way are likely to do so because of assortative mating.
3. A measure of parental investment in a child is whether they are prepared to pay for his education privately rather than send him to a state school.

4. Therefore couples who both vote the same way are more likely to send their child to a private school than are couples who vote differently.

I wish this were a joke, but it is not. I heard the report of these findings presented with all solemnity at a meeting of the prestigious Association for the Study of Animal Behaviour at London Zoo, when two human sociobiologists[35] reported that parents who both voted Conservative were more likely to send their child to a private school than if one parent voted Conservative and the other Labour; QED. Other than myself, I don't think anyone at that meeting found at all startling the claim that this was appropriate evidence for the kin selection mechanism. Haldane had been dead for many years by the time his offhand remark found such a bathetic route into scientific discourse, but granted his strongly held socialist[36] as well as scientific views, I doubt that he would have been amused.

One poor example doesn't demolish a theoretical structure, and, as I say, granted the gene's-eye view of the world on which it is based, the syllogism of kin selection is unassailable. The question is not whether it occurs, but whether it, together with the similarly gene's-eye view of the origins and maintenance of social organization that it implies, is sufficient to account for the rich varieties of behaviour that we observe in both the human and non-human animal worlds. It is at this point that it becomes more than just universal Darwinism; it is, as I shall argue, universal ultra-Darwinism.

SPECIATION

Now to Darwin's third great problem, that of how new species come into being. The adaptationist account makes it clear how species can get better at doing their thing, as in the peppered moth example, and can even develop quite subtly interactive forms of social behaviour. Such evolutionary processes could obviously modify a species over time to such a degree that its members would no longer be able to reproduce with their ancestors (if the ancestors could somehow be brought back to life). In this sense, species can gradually be transformed through processes of natural selection steadily tracking environmental change. But this still doesn't explain how natural selection alone, in the purely neo-Darwinian sense and on the basis of the mechanisms I have discussed, can result in one pre-existing species splitting into two. For this, additional mechanisms are required. Evolutionary biologists have wrestled with this paradox ever since Darwin himself recognized the problem. But Darwin's own observations of the bird species found on the Galapagos Islands also provided one of the best examples of how speciation might have occurred. When he visited these islands off the Pacific coast of Ecuador during his *Beagle* voyage, he noted that they boasted a rich bird life, and shot and collected many samples. When, back in London, he tried to identify them, he eventually concluded that the collection was made up of some 12 different but closely related species of finch, each largely confined to a single island of the Galapagos group. More modern counts put the number of distinct species at 13 or 14. Some are ground-living, some live in trees; some eat insects, others are vegetarian, living on seeds or cactus. One is wood-boring. Each species has a characteristically differently shaped beak, well adapted to the specific food and lifestyle it has adopted (Figure 7.2).

Neither the degree of adaptation of each species to their individual island conditions nor the overall similarities of the birds could be ignored. Darwin was forced to the conclusion that the island finches all originated from the same mainland species, members of which had either flown or been blown out to sea and eventually colonized each island. On each island, variations more adapted to the specific conditions there, and especially to specific potential sources of food, would spread in the population, and as there could be no cross-breeding between the birds on their separate islands, over time the accumulated variations became so great as to constitute a new and different species on each island.[37]

Thus more than just natural selection is required for a single species to split: there must also be a period of reproductive separation between two populations of the species. The Galapagos Islands provided such a separation mechanism, and in general it is now assumed that the

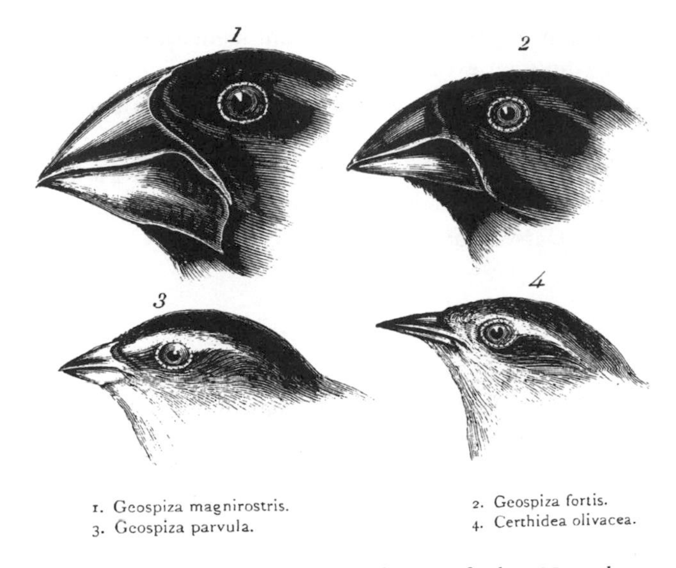

Figure 7.2 Four of Darwin's Galapagos finches. Note the different sizes and shapes of the beaks of the species shown, adapted according to each bird's particular diet.

easiest way to achieve reproductive separation is a degree of geographical isolation; barriers such as mountain ranges, deserts and seas are all potential separators.

Provided there is always genetic variation within a population, if a small number of individuals make it across the geographical barrier to their new potential home, they will represent but a subset of the original population, and, reproducing in their new environment, will rapidly begin to diverge from it even in the absence of strongly differing environmental pressures. Within relatively few generations the character of each population will differ enough to make fertile mating between them increasingly difficult, and ultimately impossible. Even if the two populations then re-meet, they will do so as distinct and reproductively isolated species.

These benign *founder effects*, as they are known, may be a major means of speciation. They are surely not the only one. More catastrophic events – devastation of a local environment by climate change, fire, earthquake – or even a giant meteorite, claimed as the cause of the dinosaur extinctions – may presumably destroy so much of a population as to make it inevitable that chance variations present in the remainder will spread. But could this be all? For the orthodox neo- or ultra-Darwinian, there is nothing else available. It is the task of the next chapter to move beyond the restrictive bounds provided by such ultra-Darwinism.

NOTES

1. Daniel C. Dennett, *Darwin's Dangerous Idea.*
2. Richard W. Burkhardt, *The Spirit of System.*
3. For a systematically churlish effort to minimize Darwin's contribution, see Sven Løvtrup, *Darwinism: The Refutation of a Myth.*
4. Wilma George, *Biologist Philosopher: A Study of the Life and Writings of Alfred Russel Wallace.*
5. Adrian Desmond and James Moore, *Darwin*, p. 264.
6. James Moore, *The Post-Darwinian Controversies.*
7. John Tyler Bonner, *The Evolution of Complexity by Means of Natural Selection.*
8. Peter T. Saunders and Mae-Wan Ho, 'On the increase in complexity in evolution'.
9. John Herschel, quoted in Ernst Mayr, *One Long Argument*, p. 49.
10. Pierre Teilhard de Chardin, *The Phenomenon of Man.*
11. William H. Thorpe, *Purpose in a World of Chance.*
12. Jacques Monod, *Chance and Necessity.*
13. J. B. S. Haldane, 'The interaction of nature and nurture', quoted by Jerry Hirsch, 'A nemesis for heritability estimation'.
14. Despite the valiant efforts of some – see e.g. Jerry Hirsch, *ibid.*
15. Population in this sense is a technical term from genetics, implying a group within a species in which more or less random interbreeding occurs.
16. Nicholas J. Mackintosh (ed.), *Cyril Burt: Fraud or Framed?*
17. See e.g. T. J. Bouchard, N. L. Segal and D. T. Lykken, 'Genetic and environmental influences on special mental abilities in a sample of twins reared apart', published in *Acta genetica gemellologica*. Many of Bouchard's publications have appeared in this somewhat obscure journal, published by

the Mendel Institute of Rome, or in volumes of conference proceedings, rather than in refereed journals. This tends to make them hard to evaluate, especially as only very heavily statistically processed data, rather than primary data, is presented. I spent three months as Visiting Professor in Minneapolis a few years back, and visited Bouchard with a view to inspecting the primary data – a normally accepted academic procedure – only to be told that he was bound by codes of confidentiality not to reveal it. Others have apparently had the same experience, which makes it very difficult to evaluate the significance of his claims.

18. These figures come from a book by a convinced hereditarian, declared racial supremacist and former student of Hans Eysenck, J. Philippe Rushton: *Race, Evolution and Behavior*, p. 83.

19. Neo-Nazi, fascist and racist groups regularly claim both that racism and xenophobia are 'in our selfish genes' and that science has 'proved' that blacks are genetically inferior to whites in terms of IQ etc. For references see for instance the chapter 'Less than human nature: Biology and the New Right' in my 1987 essay collection *Molecules and Minds*. See also the claims made by self-styled 'scientific racist' Christopher Brand in his 1996 book *The g Factor*, which John Wiley published but then withdrew.

20. Thomas Bouchard, 'Experience producing drive theory . . .'

21. *The Bell Curve*, by Richard J. Herrnstein and Charles Murray, is perhaps the most striking current example.

22. Richard Dawkins, *River out of Eden*, p. 77.

23. This claim may need to be modified in the light of recent evidence for genetic homologies among different light detecting systems, which may indicate a common origin – an observation which in fact would probably suit Dawkins' argument better in some ways.

24. H. B. D. Kettlewell, 'Selection experiments on industrial melanism in the lepidoptera'.

25. Adam Kuper, *The Chosen Primate*.

26. D. L. Perrett, K. A. May and S. Yoshikawa, 'Facial shape and judgements of female attractiveness'. For the argument taken to its popular conclusion, see Matt Ridley, *The Red Queen*.

27. David Concar, 'Sex and the symmetrical body'.

28. Robin Baker, *Sperm Wars*.

29. V. C. Wynne-Edwards, *Animal Dispersion in Relation to Social Behaviour*.

30. William D. Hamilton, 'The genetical evolution of social behaviour'.

31. Edward O. Wilson, *Sociobiology: The New Synthesis*.

32. Dawkins, *The Selfish Gene*.

33. Robert Trivers, 'The evolution of reciprocal altruism'.

34. See, among other sociobiological popularizers, David Barash, *Sociobiology: The Whisperings Within.*

35. The paper was delivered at a conference of the Association for the Study of Animal Behaviour, but seems never to have been published – perhaps not surprisingly!

36. Haldane was a Marxist and a member of the British Communist Party for many years, allowing his membership to lapse in the late 1940s, at the time of the Lysenko affair in the Soviet Union.

37. For the history of the finches and its implication, see Jonathan Weiner, *The Beak of the Finch.*

8

Beyond Ultra-Darwinism

You can drop a mouse down a thousand-yard mine shaft; and, on arriving at the bottom, it gets a slight shock and walks away. A rat is killed, a man is broken, a horse splashes.

J. B. S. Haldane, *On Being the Right Size and Other Essays*

ULTRA-DARWINISM

Let me begin by summarizing what I describe as the ultra-Darwinist position. I shall put it in its most direct and unvarnished form, although I am aware that in so doing I run the risk of caricature. None the less, there is a clear heuristic value to being so blunt, as it will provide the backdrop to my account of those positions that contest the space occupied by the ultra-Darwinists. As I see it, ultra-Darwinism has a metaphysical foundation upon which are constructed two premises. The metaphysical foundation is straightforward: the purpose (*telos*) of life is reproduction, reproduction of the genes embedded in the 'lumbering robots' which constitute living organisms. This goal can be expressed with varying degrees of sophistication, most bluntly, perhaps, as in a *Time* magazine cover story in 1995: '. . . getting genes into the next generation was, for better or worse, the criterion by which the human mind was designed'.[1] Every living process is therefore in some way directed towards this grand goal.

The two premises follow from this foundational metaphysic. One premise describes an object, the second a process. The first states that the unit of life, that which is the minimal life form, is an individual gene. These genes are not the genes of the molecular biologists, strands of DNA intertwined with histones and in dynamic interaction with cellular components to create the fluid genome. Rather they are a bit like atoms were before the days of nuclear physics: hard, impenetrable and indivisible billiard-balls, whose mode of interaction with one another and with their surrounding medium is limited to a collision followed by a bounce. The sole activity and telos of these genes is to create the conditions for their own replication – that is, to ensure the synthesis of identical copies of themselves – packaged either in the form of a dividing cell or of a reproducing organism. The genes direct the development and physiological function of the organism. How they function may be modified by random mutation, but nothing in the life experience of the body they inhabit and control can feed back to them in such a way as to improve the copies of themselves they pass on to the next generation. To repeat: 'once "information" has passed into the protein it *cannot get out again*'.

The second premise describes a process, that of adaptation. Every observable aspect of the phenotype of an organism – its biochemistry, its form, its behaviour – is in some way adaptive. It has been selected for by the honing force of natural selection, which has ruthlessly carved away any aspect of the phenotype which is less fit – that is, less able to provide the survival machine which will enable genes in due course to copy themselves. Of course, this statement must immediately be modified, for although most deleterious mutations are eliminated, there are some which, despite resulting in inferior phenotypes, may somehow be preserved in the population, perhaps because they confer some unexpected advantage.

There are cases of this sort, the best known being the abnormal haemoglobin in sickle-cell anaemia, coded for by a recessive gene. Although *homozygotes* (those with two copies of the abnormal gene, one from each parent) for the condition are at a severe disadvantage, it is believed that *heterozygotes* (those with only one copy) receive some protection against malaria. The gene mutation, arising in a human population living in a malaria-prone environment, is therefore preserved, despite the problems faced by those that are homozygous for it. But it has to be said that this argument is often stretched beyond the bounds of credulity, as when Wilson argued that 'genes for homosexuality' could be preserved in the population because they might make their bearers particularly supportive in the bringing up of children of their kin, as aunts or uncles, and thus indirectly help perpetuate their own genes, even though those who carried them were less likely to have children. (Wilson defends his hypothesis from any type of evidential refutation by claiming that such genes, if they exist 'are almost certainly incomplete in penetrance and variable in expressivity'.[2])

There is of course no empirical evidence for Wilson's propositions – it is not clear that homosexual men and women necessarily have fewer children than heterosexuals, as few fall exclusively into either category, nor is it apparent that gays or lesbians provide particularly good support in rearing their siblings' children. Nor, to my knowledge, despite some attempts to demonstrate it in captive populations, is there any evidence that homosexual behaviour is widely displayed among non-human social animals as it should be if the argument of genetic advantage holds. An alternative hypothesis for the genetic origins of homosexuality, based entirely on speculation but none the less published in a minor but respectable biological journal, is that heterozygotes for 'the homosexual gene' might be at some selective advantage, rather as with sickle-cell anaemia, because their sperm are in some way 'fitter' and therefore more effective in competing with sperm that do not carry the gene.[3] As I, Dick Lewontin and Leo Kamin pointed out in *Not in Our Genes*, the sensation one gets when reading such stuff, of being a voyeur at one's fellow biologists' more outlandish sexual fantasies, is sometimes overwhelming.

THE GENETIC METAPHYSIC

There are two features of this metaphysic which link it to philosophical positions which long predate it. The first combines the views of the moral and political philosopher Thomas Hobbes with those of the economist Adam Smith. Hobbes, as is well known, saw human life as nasty, brutish and short – a war of all against all, preventable only by State control. So it is with the competitive, selfish genes postulated by ultra-Darwinism. But how, under this condition of ruthless competition, can one achieve anything like a harmoniously functioning organism? To account for the seemingly integrated workings of a competitive society, Smith invoked 'the invisible hand of the Market' which, when each individual acted in his or her own perfect competitive self-interest, would result in a society seemingly unregulated but none the less functioning in the best interests of all. So it is with the selfish genes of ultra-Darwinism, producing higher-level order – even cooperation – from competitive individualism.

The second feature of the ultra-Darwinian metaphysic is its restatement in scientific form of one of the several Christian theologies: preformationism. We are the product of our genes, themselves the product of previous genes, themselves the products of . . . stretching back, if not to Adam, then at least to mitochondrial Eve and her un-named partner, the putative great, great ancestress of us all. We are but the carriers of this precious genetic fluid. Our task is to preserve and transmit it in our turn; but although it shapes us, we are incapable of modifying it – we merely live out its genetic instructions, albeit in an environment not (entirely) of its own choosing. The theological message is clear. However, as will become apparent, this unit-gene centred world-view soon leads to problems, just as does the concept of a genetic market economy.

Obviously, the genes being preserved in this way are the sociobiologists' rather than the biochemists' genes, as represented in Table 5.1 (page 127). Biochemists' genes are DNA molecules, metabolically engaged in all the processes of transcription and translation. When the double helix unwinds and a second complementary nucleotide strand is enzymically synthesized to match each of the two helices, the resulting molecules are each half new. In the many cell divisions that occur between conception and adulthood the 'original' DNA transmitted from parents to offspring will have been diluted millions of times over by the newly synthesized molecules, and thus will be present in homeopathic quantities even if it had avoided the risk of being degraded in the interval. When the new adult mates and generates offspring, the odds against the DNA they acquire containing any of the molecules present at its parents' conception are unimaginably large.

What, then, is meant by the preservation and transmission of genes? Clearly not the persistence of the DNA molecules themselves, but rather the replication of form, distinct from composition. There is no chemical or physical continuity. It is in this sense somewhat analogous to the capacity of the form of an organism to persist despite the fact that every molecule and (almost) every cell in its body is being continually degraded and replaced by others, more or less identical. The metaphor of replication masks the biochemical processes involved. To speak, even metaphorically, as if the DNA had an 'interest' in its own accurate replication is to traduce the complexity of the biochemical processes, to introduce a metaphysical notion of 'the gene' which the chemical structures of DNA themselves belie.

REBELLING AGAINST TYRANNICAL REPLICATORS

It follows from the ultra-Darwinian metaphysic and associated premises that the prime function of every living organism, obeying the instructions of its genes, is to maximize its inclusive fitness – that is, to ensure the greatest possible spread of its own and its close relatives' genes in succeeding generations. This does raise a particular paradox with us humans. Even the most hard-line of ultra-Darwinists manifestly do not conduct their own lives according to their own ultra-Darwinian precepts, by sparing no effort to maximize their inclusive fitness. How do they account for this apparent genetic failure on their part? Let Dawkins explain:[4]

> We are built as gene machines . . . but we have the power to turn against our creators. We, alone on earth, can rebel against the tyranny of the selfish replicators.

Whence this power? According to Wilson, it arises because, although genes 'hold' culture, they do so 'on a leash'.[5] There is something fundamentally unsatisfactory about this argument. Either we, like all other living forms, are the products of our genes, or we are not. If we are, it must be that our genes are not merely selfish but also rebellious, building the phenotypic structures that give our brains and culture the power to contradict the orders of some of the other replicators embedded in every cell of our bodies. And as our brains are the product of evolution and did not fall independently from the sky, nor were they generated by a highly un-Darwinian massive mutational leap, there must presumably be at least a germ of rebelliousness in the genes of our near evolutionary neighbours too. The selfish genetic imperative is hoist with its own petard.

If, on the other hand, it is not our genes that are rebellious, what other options are available? Dawkins never says, but implicit in his argument is that somewhere there is some non-material, non-genetic force moulding our behaviour. This is dangerously close to Descartes, with his mind or soul in the pineal gland directing the mere mechanism which constitutes the body. For Descartes, non-human animals are of course mere machines, and I suspect that he would have been perfectly at home with Dawkinsology. Thus, despite Dawkins' passionately explicit claims to atheism and expressed hostility to religion, the charge against him (and his fellow ultra-Darwinians) is that they fail to carry their own genetic argument to its logical conclusion, which is that it has to be our genes that make us 'free' and 'rebels', and provide us with the plasticity enabling us to modulate our culture, on however long a leash it is held. As a result, ultra-Darwinists re-import dualism – a dualism which is central to Christian theology, but absent from that of other religions, such as Buddhism or Confucianism – by the back door. Brian Goodwin has somewhat mischievously pointed out that this ultra-Darwinian syllogism, with its final concession to salvation through good works, is remarkably similar to Christian fall and redemption myths,[6] despite the avowed antireligious sentiments of its proponents. As Shakespeare put it, 'There's a destiny doth shape our ends, rough hew them how we will.' For 'destiny', read 'genes'.

THE CASE AGAINST ULTRA-DARWINISM

It is time to look in more detail at the scientific case against ultra-Darwinism. This rests on the following claims:

1. The individual gene is not the only level at which selection occurs.
2. Natural selection is not the only force driving evolutionary change.
3. Organisms are not indefinitely flexible to change; selection is '*table d'hôte*' and not '*à la carte*'.
4. Organisms are not mere passive responders to selective forces, but active players in their own destiny.

I shall examine each of these propositions in turn. Let me be clear that my critique is in no way directed against the *fact* of evolution among the organisms that inhabit our planet, nor against the mechanisms of natural selection that Darwin himself proposed. There will be no comfort for creationism, fundamentalist religions or New Age mysticism here. My principal target is the dogmatic gene's-eye view of the world that ultra-Darwinism offers. There is more, much more, to life, and to evolutionary change, than is dreamt of in the ultra-Darwinists' philosophy. As will become apparent as the argument unfolds, their position is tenable only on the assumption, which the previous chapters have challenged, of a direct and relatively unmodifiable line between gene and adult phenotype. There is no room within the model for the processes of development or for the internal physiological processes which constitute the organism. This debate will lead us back, in the next chapter, to a consideration of the metaphysical foundation of ultra-Darwinism in its approach to the origins of life itself.

LEVELS OF SELECTION: GENES OR GENOMES?

It is clear from the description of the modern concept of genes as relatively indeterminate sections of DNA, interspersed with non-coding regions, capable of multiple forms of processing, editing and reading before the proteins for which they code are ultimately dispatched to fulfil their several cellular functions, that the ultra-Darwinists' metaphysical concept of genes as hard, impenetrable and isolated units cannot be correct. Any individual gene can be expressed only against the background of the whole of the rest of the genome. Genes produce gene products which in turn influence other genes, switching them on and off, modulating their activity and function. If selection ultimately determines whether a particular gene survives or not, it can do so only in context. To go back to an example in the previous chapter, a gene 'for' making antelopes run faster will not be selected in the context of another gene which also ensures that they spend longer in a vulnerable infant state before maturing into fast-running adults.

But even to use this language is to fall into the trap set by an ultra-Darwinism in which genes are still the inferred entities conferring phenotypic properties, rather than the material objects of biochemical investigation. That is, 'a' gene is selected only if it results in a selectable phenotypic change – yet what is required to produce such a change is not one but many actual biochemical gene-sized lengths of DNA. One gene alone will not produce the wide range of changes, in body size, metabolism, bone structure, and so on which may be required to produce a faster-running antelope.

In fact, this was recognized long before present-day molecular biology, by the third of the geneticists whose 'modern synthesis' united Mendelism and Darwinism in the 1930s, the American population biologist Sewall Wright. Where Fisher and Haldane had considered the properties of individual genes, Wright insisted that the whole genome needed to be taken into consideration. Fisher and Haldane's approach was derided as 'beanbag' genetics precisely because it depended for its mathematics on the assumption that each gene was an isolated unit which could be shaken, shuffled and selected like one bean in a beanbag independently of all others. The insistence on the whole genome, and the study of evolution in action in naturally occurring populations, rather than the controlled experimental plots at Rothamsted, also characterized the flowering of genetics in the young Soviet Union until, towards the end of the 1930s, the science was effectively destroyed by Stalin and his protégé Trofim Lysenko.[7] Theodosius Dobzhansky, who left Russia for the United States at the end of the 1920s, became heir to both the early Soviet and the Sewall Wright traditions, helping to ensure that (except among the behaviour geneticists and psychometricians, who think in beanbags whichever side of the Atlantic they are located) the contrast between beanbag and genomic thinking has characterized the distinctive traditions of British versus American population genetics ever since.[8]

LEVELS OF SELECTION: GENES, CELLS AND DEVELOPMENT

On ultra-Darwinian, or even Weismannian principles, the genome you inherit is the one – granted the shuffling that goes on during sex – that you pass on to your offspring via your own genes in your gametes (or germ-cells). Materials for change are available only by courtesy of mutation in these genes. Is this entirely true? Is there any way in which an individual's lifetime experience could affect the genes – that is, could Lamarckian mechanisms apply to at least some aspect of evolution? This proposition, always attractive to anti-Darwinians, of course runs counter to Crick's Central Dogma, and strictly speaking the answer is surely 'No'. Yet there is mounting experimental evidence that among bacteria there can indeed be adaptive mutations – that is, mutations in some sense directed by environmental conditions, so that they can occur under circumstances where they might contribute to the survival of the organism much more frequently than might be expected on a purely random basis.[9]

The situation is more complex in multicellular eukaryotes, where replication entails not merely sex but, crucially, development. There are aspects of the developmental process which seem to leave some scope for adaptive, rather than chance mutations. Developmental biologists have wrestled with this question for decades. In a sense, the argument goes back to Darwin's resistance to the suggestion that evolution could proceed by leaps – *saltations* – much to the distress of many of his otherwise enthusiastic followers, such as Francis

Galton. In the 1930s, the evolutionary geneticist Richard Goldschmidt suggested that significant adaptive changes could occur by a process of pre-adaptation, the creation of what he called 'hopeful monsters' equipped with the mutations necessary for some appropriate substantial change, and awaiting the appropriate environmental circumstances to make the leap.

Goldschmidt's ideas have never won acceptance among evolutionists or geneticists, and an alternative way out of the dilemma was proposed by Conrad (Hal) Waddington, an Edinburgh-based theoretical biologist much influenced by the work of the Cambridge Theoretical Biology Club of the 1930s. He argued that developmental processes in multicellular organisms could help both direct and, as he put it, 'canalize', potentially favourable mutations. Waddington's ideas, focused through the organization of a series of highly influential conferences and published volumes through the 1960s, helped shape a new developmental perspective on evolutionary change.[10] Empirical evidence for such processes is hard to come by, but the Harvard developmental biologist John Tyler Bonner[11] has built on Waddington's ideas by pointing out that Weismann's barrier cannot be as fixed as ultra-Darwinism implies, for two main reasons. The first is rather subtle, and applies only to plants and a relatively limited group of small invertebrate organisms; the second is universal.

To deal with the subtle case first: the Weismannian principle is that, from the earliest stages of development, the gametes (Weismann's germplasm) are sequestered from the rest of the body (the *soma*), and hence cannot be influenced by factors which affect it. Bonner points out that, while this is generally true for more complex animals (meaning animals with greater numbers of distinct types of body cell), it is not true for plants, or for less complex animals such as the tiny pond-dwelling hydra. Like plant cells, the cells of the hydra retain the capacity either to differentiate into somatic cells, or to become sequestered as gametes, or to remain totipotent. Those cells which remain totipotent retain the prospect of becoming gametes after an indefinite number of cell divisions – and this means that any genetic variation occurring during those divisions will be heritable (Figure 8.1). Weismann's barrier does not apply.

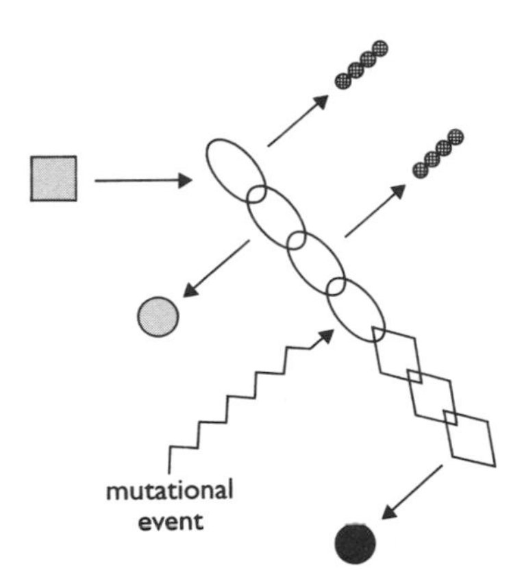

Figure 8.1 Totipotency: how genetic variation can occur in cells beyond Weissmann's 'germplasm', as proposed by John Tyler Bonner. The original gamete (grey square) gives rise to stem cells (open ellipse) which can differentiate into functional somatic cells (hatched circles) or into gametes (grey circles). A mutation occurring in a stem cell (open lozenge) can thus give rise to a mutant germ line cell (black circle).

Important as this argument is in breaking the dead grip of Weismannism, until recently it seemed not to apply to more complex animals. This assumption has been made increasingly doubtful by recent advances in gene technology, however. In 1996 an Edinburgh-based team directed by Ian Wilmut succeeded in cloning sheep from embryonic cells, and the following year announced in a paper in *Nature*[12] which attracted world-wide attention, that they had performed the same operation using DNA extracted from cells obtained from the udder of an adult sheep. The ethical issues and media concern raised by this experiment are not of direct concern to me here; the relevant point from the perspective of the argument in this chapter is that adult sheep DNA and the cells from which it is derived remain totipotent. Weismann's barrier is well and truly breached.

However, there is another, more universal issue to which Bonner points, drawing on earlier insights by the Scottish biologist and philosopher Lancelot Law Whyte,[13] who described what he called 'internal factors' in evolution. During development, originally totipotent cells divide, become determined and migrate to appropriate positions within the developing embryo. Migration, as discussed earlier, depends on complex factors including internal features of the cells themselves, the presence of appropriate tissues or surfaces over which they can move, information arriving in the form of secreted chemicals from their neighbours sharing the migratory journey, and 'trophic factors' diffusing out from their target organs and signalling the directions in which the migrant is to move.

This process has the consequence that a type of competitive/selective mechanism operates between cells within the developing organism itself. Many more cells are generated during embryogenesis than ultimately survive. Those cells that fail to make the migratory journey adequately, or arrive too late, are lost; they will leave no progeny, no daughter cells. What determines success or failure in this migratory journey? Cooperative relations both among the migrating cells and between them and their target organs through their trophic secretions will be part of the mechanism. Contingency – sheer accident – may be another. But there may also be variations between the cells, making selection possible in the classical Darwinian sense, as already discussed in the context of Edelman's selection hypothesis. This developmental process, demanding what Bonner calls 'sound rules of construction',[14] must itself be subject to strong selection pressure, but will also constrain the final outcome, the mature, reproductively competent phenotype. There should be nothing surprising about this. Any large organization has simultaneously to act as a coherent unit in its relations with the outside world, in cooperation and competition with its peers, while at the same time serving as the cockpit for the internal power struggles, the jostling for position, the personal ambitions, of its component members. Once again, this complexity is lost to the one-dimensional world of ultra-Darwinism.

LEVELS OF SELECTION: GENES AND PHENOTYPES

So selection acts on genes, on genomes and on cells, notably during development. But for multicellular organisms it is ultimately the organism as an integral unit that will or will not reproduce and dispatch copies of its genes to subsequent generations. So natural selection in the sense that Darwin originally conceived it can operate only through the actions and properties of the entire organism, its phenotype. For ultra-Darwinists that is not a problem: the phenotype is merely a proxy for the genes it contains, the gene's way of making copies of itself. But this implies a direct relationship, one-to-one, between gene and phenotype, which of course is exactly how the ultra-Darwinists speak, and is why, for molecular biologists, organisms virtually cease to exist except as probes for the study of genes. Read almost any sentence of Dawkins' *River out of Eden*, and you will find this rhetorical proposition screaming out at you.

Such a claim ignores development and the complex processes whereby genes active only at one point in time and space during an organism's lifeline are switched on and off, and the fact that the survival of any gene to the point at which the organism is mature enough to reproduce depends upon the 'goodwill' of other genes. It ignores the presence of the so-called 'selfish' DNA, the seemingly genetically meaningless introns (described in Chapter 5) which comprise the bulk of the DNA in the genome. If copies of all this DNA can be carried along, generation after generation, without any apparent phenotypic effect at the level of the organism, then the cause-and-effect linearity which ultra-Darwinism offers for gene–phenotype relationships is massively violated. The only 'phenotype' of such 'selfish' genes is the actual DNA that constitutes them. Thus 98 per cent of the DNA in the human genome is without phenotypic significance at the level of the individual (though it might be interesting, using modern genetic engineering techniques, to see what would happen if one were to construct a chromosome minus the apparently redundant introns).

But this speculation apart, the claim of specificity ignores the reciprocity of gene–environment relationships: for example, particular phenotypes can be the result either of the presence of particular genes or of particular environments. And it ignores the fact that, as I have constantly emphasized throughout this book, whereas there is only one level at which the gene can be described, the term 'environment' is multi-layered, ranging from the intracellular to the global. Thus, while a small proportion of cases of breast cancer or Alzheimer's disease is attributable to particular 'major' genes, in most such cases these genes are absent. Instead, unspecified and largely unknown environmental factors result in the same end-point. Most geneticists accept that this is so, although increasingly common now are hard-line claims that what appear to be 'environmental' effects are in fact attributable to 'genetic' risk factors, with many 'minor' genes, each with a small potential to affect the final outcome, acting in synergy either with each other or with environmental factors to produce the disease.[15] Those who offer such 'minor gene' alternatives still insist on the primacy of genetic explanations, accepting rather grudgingly the possibility of environmental determinants. To emphasize the subordination of the environmental to the genetic explanation, such conditions are called 'phenocopies'. As the obvious genetic causes in these cases are in the minority, I prefer to reverse the terminology and refer instead to 'genocopies'. That little ideological spat apart, the point is clear: there is not and cannot be a simple one-to-one relationship between any given gene and the phenotypic expression at the level of the mature, fully developed organism. The multiple layers of interaction and levels of complexity which separate DNA strands from lifelines ensures this.

One of the major causes of mutation in DNA strands is the level of cosmic rays (high-energy subatomic particles entering the Earth's atmosphere from outer space), which provides a more or less steady source of variation. There is a good deal of technical debate in the specialist literature about whether and why the mutation rate in different species subject to the same background level of cosmic rays may vary, but that need not concern us here. The question of relevance to the issue of gene–phenotype relations is whether, granted a reasonably constant mutation rate in DNA, there is a corresponding change in phenotypic expression, which a one-to-one gene–phenotype relationship might imply. Of course, in part this depends on how one defines the level of the phenotype one is studying. If the gene's phenotype is the DNA itself, then the answer is obviously that there is such a change. But if the phenotype is the entire organism, and if in particular the mutations are the source of phenotypic variation on which selection can act, then the answer is much more complicated. If lifelines depend on order at many levels of organization, as I have argued in Chapter 6, then at each level damping processes will occur so as to minimize the effects of minor variations, unless and until those variations are of a magnitude to drive the autopoietic structure into a different stable state. Two types of empirical observation bear this out.

The first became apparent when, during the 1970s, certain new techniques of protein separation began to be applied to problems in population genetics. Proteins which differ even subtly from one another in molecular weight and electrical charge can be separated rather simply by gel electrophoresis, the process described in Chapter 3. Basically, you will recall, this involves making a thin slab of jelly (gel) out of starch (or, more commonly these days, polyacrylamide), putting a drop of a protein containing solution at one end of the gel, and passing an electric current through the length of the gel. Proteins are pushed along the gel at a rate which depends on their charge and molecular weight. If the current is switched off after a few hours, and the gel is submerged in a dye solution or mix of substrates for particular enzymes which stain the different protein fractions, the fractions become visible, strung out along the gel like a freeze-frame of sprinters on a running track.

In 1966 Richard Lewontin, a geneticist reared in the Dobzhansky tradition, applied this technique to the proteins derived from *Drosophila* populations, and discovered that there were considerable variations in the numbers and distribution of *isoenzymes* (enzymes of different protein structure but which catalyse identical reactions). There were variations not only between populations recovered from different regions of North America, but also within any given population of flies.[16] There is a great deal of hidden variation, then, even within a population which seems to contain rather similar phenotypes.

The different isoenzymes may each be the product of a particular allele, or they may be generated by different splicing and editing procedures from the same gene-sized length of DNA, but their very presence immediately raises doubts about simplistic ideas of one-to-one gene–phenotype relationships. The discovery of this phenotypic diversity made population geneticists and evolutionary biologists ask whether particular combinations of isoenzymes provide a selective advantage to their possessor – in which case the variations are adaptive and provide the material on which natural selection can operate – or whether they are contingent, historical accidents which are essentially selectively neutral. Diversity would be maintained simply because it has arisen, rather like the persistence of 'selfish DNA' among the introns. I shall return to this question in due course.

A second query over the tight coupling of genotype and phenotype began to be raised by the palaeontologists Stephen Jay Gould and Niles Eldredge. They studied fossils of trilobites, species once common but now extinct. Fossilized trilobites can be found in rock strata deposited over an extraordinarily long period of evolutionary history, no less than sixty million years. Gould and Eldredge pointed to the apparent paradox that, despite what must presumably to have been a steady mutation rate in the trilobite DNA over the whole of this period, the fossilized body forms of the organisms are remarkably constant: phenotypic stability imposed upon genetic variability. Where there had been phenotypic change, it seemed to have occurred in bursts, over relatively short periods of geological time. Since the nineteenth century, orthodox Darwinism has stressed gradualism in evolutionary change. By contrast, Gould and Eldredge's theory emphasized phenotypic stability over long periods, alternating with periods of intense phenotypic change. They called this *punctuated equilibrium*.[17]

Despite the respect with which its authors are held by most of the palaeontological community, the thesis of punctuated equilibrium could scarcely be said to have won universal acceptance among evolutionary biologists. Maynard Smith, for instance, has pointed out that whether one regards a period of evolutionary change as brief or not depends on a geologist's time-perspective. For palaeontologists a period of a million years is little more than the blinking of an eye. On this time-scale, Gould and Eldredge's punctuated equilibrium may not be such a heresy after all.[18] Furthermore, as the fossil record tells us primarily about preserved hard structures and not about either proteins or behaviours, we cannot know how the lifestyle of the trilobites may have changed despite their seemingly invariant structures. However, within the framework of multi-level order developed in earlier chapters, the suggestion that genetic variation can be damped, rendered essentially neutral, until such time as it accumulates sufficiently to tip the next generations of organisms into new stable states, seems perfectly credible.

LEVELS OF SELECTION: GENES, POPULATIONS AND SPECIES

The arguments advanced against ultra-Darwinism in the preceding sections have all focused on the organism. They may be summarized as follows. Because genes are in genomes are in developing cells are in multicellular organisms, the relationship between gene *A* and phenotype A is non-linear, and each level of organization, and indeed each moment during the developing trajectory of an individual organism's lifeline, offers an opportunity for selection to act. To paraphrase Wilson's view – and even accepting the kind of simplistic causal chain from which I have been at pains to dissociate myself – genes hold phenotypes, and not merely culture, on a long leash.

But there is more. Organisms do not exist in isolation, but in populations – populations in ecological communities in which many hundreds or thousands of different species are locked into relationships which may be competitive or cooperative. Ecologists define species as occupying *niches*, sites in which they can make a living because of their particular specialisms, like the different finches on each of the Galapagos Islands. But each species' niche is defined as a space shaped by all the other species with which it comes into contact. Two species may be predator or prey for each other, they may be parasite and host, they may live mutualistically, each necessarily dependent on the other, or they may merely be commensal – sharing the same space.

But as all species are evolving, the evolution of any one is shaped and constrained by that of many others.

Within any population, and perfectly acceptably within the Darwinian framework, it is possible to conceive of the adaptive coexistence of members with very different phenotypes. Indeed, it is John Maynard Smith, doyen of Darwinism, who has argued this idea most elegantly. It is clear that populations are able to maintain relatively stable ratios of organisms possessing both different genotypes and different phenotypes. The most obvious example is the existence of approximately equal members of each of the two sexes. There are convincing mathematical reasons, derived initially by Fisher and expanded by Maynard Smith, why this should be so, despite the very different relative contributions to the reproductive process that each makes, which might suggest the possibility – at least in mammals – of getting by with far fewer males than females.[19]

However, there are also less obvious examples, drawn from aspects of social behaviour, which Maynard Smith has attempted to model using the mathematics of game theory. This is a theory which describes the outcomes of potential strategies each of two players may adopt in confronting each other in games subject to simple rules, like noughts-and-crosses or rock–scissors–paper (ick-ack-ock). Maynard Smith uses this approach to consider rather abstract models of animal social conflict. For instance, animal populations may contain 'hawks' which fight with increasing vigour until wounded or their opponent retreats, and 'doves' which retreat from such a conflict before being injured. The algebra predicts that in populations consisting entirely of doves, a hawk mutant will be successful, and hence increase its numbers; similarly, in a population composed entirely of hawks, dove mutants will be successful. The point at which hawks and doves balance one another depends on the arbitrary numbers assigned to the algebraic formulations, but the essential outcome is a relatively stable ratio of hawks to doves. This, says Maynard Smith, is an evolutionary stable strategy.[20]

One may argue that such abstract examples are far removed from real life, but they do demonstrate, albeit simplistically, that balance can exist between members of a species showing very different types of behaviour. Evolutionary stable strategies mean that in socially living animals, selective processes can result in a mutually evolving population. For such a population it makes no sense to consider the selective advantage of one gene or genotype except against the background of the entire population of genotypes within that population – in the same way that, as I argued earlier, it makes no sense to consider the selective advantage of a single gene except against the background of the entire genome of the individual organism. Thus we have yet another level of selection: that of the population as a whole. To cling to 'the gene' as the sole unit and level of selection under these circumstances, as Maynard Smith and the ultra-Darwinists do, seems perverse, a point made with great force by the successor to Dobzhansky's mantle, Ernst Mayr, in his *magnum opus* on diversity, evolution and inheritance.[21] The consequence of this logic is in essence that group selection mechanisms, although probably not in the form in which they were originally conceived, are seriously back on the agenda of mainstream evolutionary theory.[22]

But this co-evolutionary argument need not be confined to members of an individual species, for it must also reflect the relationships between members of different species which share their living space. Some of these are obvious. Consider the mutual interactions between plants and the insects which pollinate them. Plants produce flowers, which encourage bees or other insects to settle on them. In the process, the bees collect pollen which can be transferred to the next flower on which they settle, so fertilizing it. The bees obtain a foodstuff – the nectar – and the plants get to breed. In an even more complex example of co-evolution, parasitic wasps inject their eggs into caterpillars, which as a result become paralysed while the eggs develop into wasp larvae. The wasps find their prey by homing in on volatile chemicals which they can detect over large distances. The chemicals come from the caterpillar faeces, but also, more surprisingly, from the plants on which the caterpillars feed. The plants have evolved a mechanism for secreting the chemicals to attract the wasps when they, the plants, are attacked by the caterpillars![23] This mutually advantageous system must be the product of a co-evolution in which both wasps and plants, two very different living forms, are selected more or less in parallel.

These are examples of co-evolution through the mutual cooperation of individuals within populations or between species. But taken to its logical conclusion, the relationships between antagonistic species – predator and prey, for instance – also imply co-evolution. If peppered moths are replaced by darker forms, selection pressures on the robins and hedge sparrows which prey upon them might favour forms with improved eyesight to distinguish the dark moth on the blackened bark – or they might favour forms which seek and devour alternative prey. Or both. As with so much of biology, it all depends. When the viral disease myxomatosis dramatically reduced rabbit populations in Britain in the 1950s, the populations of animals which preyed on the rabbit – foxes, badgers, stoats, weasels and buzzards – were all depleted. So were the minotaur beetle, whose larvae feed on rabbit dung pellets, the wheatear, which nests in rabbit holes, and the stone curlew, which lives on the ground cropped close by excessive rabbit grazing. Rabbit competitors, such as brown hares, on the other hand, were predicted to increase in numbers.[24]

Nothing in population relations is static, and very little is simple to forecast. How these changes in rabbit population numbers over relatively few breeding seasons altered gene frequencies in the populations is not clear, for after a few years myxomatosis-resistant strains of rabbit appeared and their numbers grew dramatically once more, until today they are at least as abundant as before the original disease struck. The point is that selection pressures are constantly changing, and evolution can do no more than track environmental change for all the species involved in the interacting web. Like the concept of homeostasis, that of 'the balance of nature', with its implicit message of unchanging stability, is profoundly mistaken. But evolutionary change follows environmental change, of course, without being able to predict it, for evolutionary forces can respond only to present circumstances, not to potential future contingencies. To isolate from this evolving web a single actor, be it gene or organism, as the unique determinant of change is as problematic as isolating a single enzyme from the metabolic web that constitutes the cell. Any such attempt at isolation is a reductionism that mistakes method for theory.

Mutualism can be taken a good deal further. What only a few years ago was a heretical idea put forward by the evolutionary biologist Lynn Margulis has now become the conventional wisdom of the textbooks. It had long been a puzzle to biochemists that mitochondria, the intracellular structures which are the principal sites of energy production within the cell, contain their own DNA, sufficient to code for a rather small number of proteins, and quite different from the DNA in the cell nucleus. Margulis was impressed by the structural similarity between mitochondria and some forms of free-living bacteria. She proposed that relatively early in the history of eukaryotic evolution a close symbiotic relationship developed between primitive eukaryotic cells, which lacked the capacity for the oxidative processes which lead to the synthesis of ATP that today characterize mitochondria, and bacteria, which had that capacity. The symbiosis she proposed culminated in the engulfing of such protomitochondrial bacteria by the eukaryotic cell, which thus acquired the capacity for oxidative metabolism. The bacteria lost their capacity for independent survival but gained the advantage of the protected internal environment of the eukaryotic cell, in which they could retain a quasi-autonomous existence.

Margulis went on to extend this idea to chloroplasts, the photosynthesizing substructures within green plant cells, and more controversially to many other subcellular structures, notably microtubules and cilia, resuscitating and developing an earlier term by describing the process of co-evolutionary development as *symbiogenesis*.[25]

In her vision, present-day multicellular organisms, both plant and animal, are the evolutionary results of a long process of closer and closer communal living between originally independent life forms. Commensality moves from close sharing of an environment to literal coexistence within the same internal space.

It is not necessary to follow Margulis all the way with her version of multicellular origins, which, despite its attractions remains to some degree speculative, in order to appreciate its implications for debates over the nature of selective processes. Evolutionary stable strategies within and between populations, whether or not they culminate in symbiogenesis, require that the 'unit of selection' now ceases to be an individual genotype or even phenotype, and becomes instead a *relationship between* genotypes and/or phenotypes. We have moved a long way from individual 'selfish genes' and their 'extended phenotypes'.

NATURAL SELECTION OF RANDOM VARIATIONS IS NOT THE ONLY FORCE DRIVING EVOLUTIONARY CHANGE

So far I have considered the nature of the unit of selection without considering the nature of selection itself. I have already pointed out that the simple Malthusian version of the Darwinian equation, selection through competition for scarce resources, can be only a partial mechanism of evolutionary change, as indeed Darwin himself well recognized; to it, whatever the level at which selection occurs, must be added sexual and kin selection, selection through founder effects, expansion of populations into novel environments or potential ecological niches as in Darwin's finches, selective predation, as in Kettlewell's moths, and co-evolution of populations and species. Furthermore, selection at any given level of the hierarchy between individual genes and ecosystems does not automatically imply selection and evolutionary change at any other. There is sufficient flexibility and redundancy within living systems to make such tight coupling unnecessary.

But is selection, at whatever level, the only motor of change? This is the second fundamental tenet of ultra-Darwinism around which great debate has centred. For it to be so, any phenotypic feature of the organism must in some way be shown to be adaptive: that is, it must confer on its owner some advantage over alternative forms in the population, thus enabling the 'Darwin machine' to operate. Do the relative concentrations of one or other of the several forms of lactate dehydrogenase, the common enzyme of energy metabolism in members of *Drosophila* populations, reflect differences in relative fitness, or are they selectively neutral? Are the variations in banding patterns found on the shells of snails within any population purely chance, or do they alter the snail's survival chances?

For the ultra-Darwinists, it is axiomatic that every such feature must represent a character which either has been selected or is available for selection. Drift or contingency are unacceptable, except as providers of the material variation on which selection can act. For its adherents, ultra-Darwinism has become a credo in which strict adaptationism replaces the 'law of higgledy-piggledy', and chance is constrained; its consequences are as predictable as the knowledge that the random processes of radioactive decay will yield isotopes whose half-life is mathematically determinable and which, brought together sufficiently closely, will result in nuclear explosion. Ultra-Darwinists seek to go beyond Darwin himself.

As usual, the arguments against such ultra-Darwinism take several forms, ranging from the empirical and molecular[26] to the theoretical and systemic. The most comprehensive critique of the adaptationist paradigm challenges ultra-Darwinism by stressing the law of higgledy-piggledy, the role of chance, of contingency, in evolution. As I have pointed out, the one thing that evolutionary processes cannot do is to anticipate environmental change, notwithstanding that any population may contain enough variability to help ensure that some variants may survive even quite significant unanticipated hazards – like the melanic forms of Kettlewell's moths. Thus, when a giant meteorite crashes into the Earth, the climate changes dramatically and the presumably otherwise well-adapted dinosaur population of the time goes extinct, leaving their territory free for the ancestors of today's mammals to flourish.

This is the argument from contingency, and it has been brilliantly expounded by Gould, this time in his book *Wonderful Life*. His account focuses on a rich fossil harvest found in a particular rock formation in British Columbia, Canada, known as the Burgess Shale. The fossils present in the shale are unique, bearing little resemblance to any presently living forms, and having body plans which seem quite weird, almost impractical (Figure 8.2). They are, as Gould puts it, of 'transcendental strangeness: *Opabibia*, with its five eyes and frontal "nozzle", *Anomalocaris*, the largest animal of its time, a fearsome

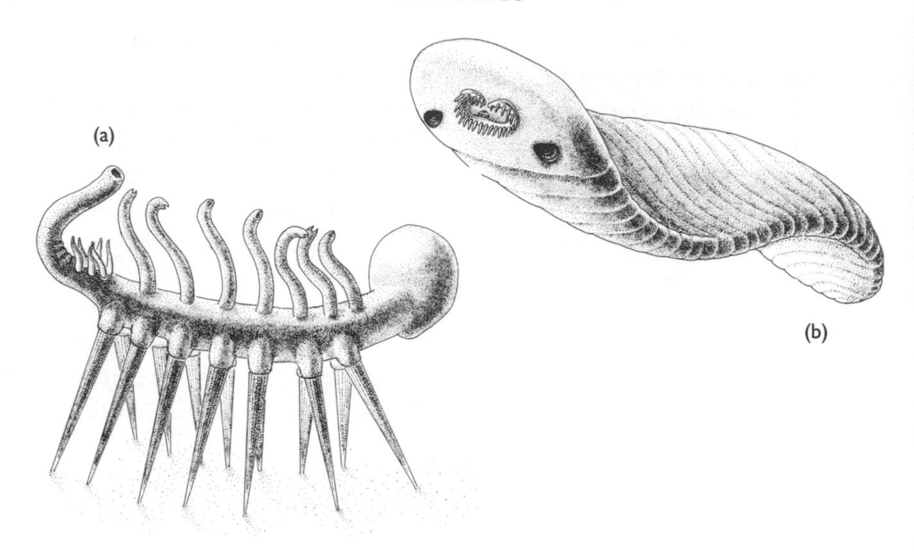

Figure 8.2 Reconstructions of two organisms preserved as fossils in the Burgess Shale: (a) Hallucigenia, *an ambiguous organism with seven pairs of struts of uncertain function, and (b)* Odontogriphus, *a flattened swimming animal with a mouth surrounded by tentacles.*

predator with a circular jaw; *Hallucigenia*, with an anatomy to match its name.'[27] All are now extinct, seemingly as a result of some catastrophe analogous to that which finished off the dinosaurs.

Had they not become extinct, mused Gould, how would the current descendants of these early life forms look? If it is indeed a mere accident that they failed to survive, then, in his frequently repeated phrase, if we could wind the tape of evolutionary history back and rerun it, it is in the highest degree unlikely that humans, or even mammals, would have evolved. Far from being the inevitable products of a strict adaptationist programme, or even the workings-out of purposive progressive evolution, we and all our works are an accident of history. Even Darwin in his dethroning of Man as the Child of God didn't go that far. Gould's argument is powerful, although it remains a matter of assertion on his part that the weird body plans of the Burgess Shale could indeed have survived, that they really were as well adapted as those more familiar organisms that did survive and are our direct though remote ancestors. On the face of it, the reconstructed Burgess Shale creatures do look a little impractical. Can it really be efficient to have five rather than two eyes, for instance? Because we simply don't know why these entire phyla went extinct, to argue that it was not a failure of adaptation, but mere contingency, is no more evidence-based than the adaptationist paradigms that are under criticism.

Which raises the final question of this section: what constitutes 'an adaptation' over which the debate about selectivity must range? The critics have characterized the adaptationist argument as 'Panglossian', after Voltaire's character Dr Pangloss, for whom everything that occurred in the world around him, even the most dramatic and seemingly negative, such as the disastrous Lisbon earthquake which so shook the so-called felicific philosophy of the eighteenth-century Enlightenment, was 'for the best in the best of all possible worlds'. The argument has taken both theoretical and ideological forms. The theoretical issues were raised within population genetics and evolutionary biology. One view was that populations were largely genetically homozygous, and such heterozygosity that existed was the result of balancing selection, as in evolutionary stable strategies. If this were the case, most variation would be adaptive, and the Panglossian paradigm would hold. All populations would be on the evolutionary track to perfectibility. The alternative view is that, to a considerable degree, chance reigns. Contingency, mutation and genetic drift result in the presence within any population of a variety of neutral mutations which may be preserved without necessarily being selectively advantageous.[28] Consider the heritable variations in the proportions of different isoenzymes present in the bloodstream among *Drosophila* populations; and heritable differences in the banding patterns on the shells of land snails. Can every such difference and its preservation within the population be explained on the grounds that it serves some function, or is it merely a matter of the perpetuation of an initial random genetic event which has no effect on survival?

The ideological issues write this theoretical dispute into human affairs. They came to a head during the first rounds of the debate over sociobiology raised by the publication of E. O. Wilson's book in 1975. As a strict adaptationist and kin selectionist, he argued that certain features of human society, which he regarded as universals – from incest taboos to male–female power relationships and individual greed, or 'indoctrinability' – were the results of selective evolutionary pressures.

Even before Wilson's book had appeared, first Dobzhansky and then Lewontin had pointed out that the assumption that there is one 'standard' wild-type of any organism, all other variants of which are deleterious mutations – the Platonic natural kind, whether of humanity or of any other species – opens the door to typological, even racist thinking. It was, however, the polemical nature of Wilson's claims that led to a re-examination of the whole adaptationist paradigm by his critics. The arguments were twofold. First, the claims for selective advantage rest on fables, rather like Rudyard Kipling's famous *Just-So Stories* about 'How the elephant got its trunk' or 'The cat which walks by itself'. There is rarely any supportive evidence for such fables, and what data there are are subject to multiple interpretations. A good example is Wilson's effort to account for homosexuality, described earlier, but the issue extends far beyond human populations, and adaptationist just-so stories are rarely without alternative explanations.

To take an example of how such an adaptationist story might work, you could argue that the red colour of blood was an adaptation – a wounded animal would bleed red and the colour could serve as a warning signal to its kin. But in fact the red colour of blood is a consequence of the iron content of its haemoglobin, and haemoglobin contains iron because it serves as a respiratory oxygen carrier. Haemoglobin is well adapted to its function as an oxygen carrier, but its red colour is surely peripheral, an irrelevant consequence of the adaptive feature for which it has presumably been selected. So although red blood may serve as a warning signal, this cannot be why blood is red, according to any of the meanings of the word 'why' in biological explanation discussed in Chapter 1. The redness is an epiphenomenon. It is interesting that in the early part of this century an American naturalist and artist did employ just this type of argument to explain why flamingoes were pink, because, when they take to flight at sunset the colour makes them hard to spot against the setting sun, and hence provides an advantage against predators. Needless to say, this adaptive hypothesis is purely fanciful.

The point is that, because 'a phenotype' may be represented at all levels from the cellular to the population, one has to be clear which feature and at which level one is choosing to tell one's adaptationist story. The most polemical characterization of the Panglossian paradigm came during a Royal Society meeting on evolution held in 1979. The early sessions had been relatively uncontroversial and self-congratulatory, cataloguing the many triumphs of the modern Darwinian synthesis. The penultimate session, however, opened with a paper by Gould and Lewontin bearing a title that has gained notoriety: 'The spandrels of San Marco and the Panglossian paradigm'.[29] (It was in fact delivered by Gould, for at that point Lewontin was going through a period of aversion to flying, and left it to his co-author to attend the meeting and present the argument; none the less the paper had all the verve and optimistic intellectual insouciance that characterized the style of both these radical critics of the conventional wisdom.)

Gould tantalized his audience of biologists by a lengthy disquisition on the architecture of the famous basilica of Venice. Look up at its vaulted roofs, Gould proclaimed, and your attention will inevitably be drawn to the sumptuously decorative mosaics that cover the panels (he called them spandrels; the more correct architectural term is pendentives) which dominate its space. An adaptationist argument will seek to explain these panels as part of an architectural design which provides surfaces at roof level on which appropriately religious messages may be inscribed (as, for example, the adaptation of the peacock's tail). And yet the pendentives are not an option, but a necessary structural element of a dome supported on arches. It is these necessary features of the design that form the pendentives; far from the roof being designed around a pendentive adaptation, it is designed around a vaulting adaptation. This, the paper concluded, is the case for many presumed adaptations, which rather than being themselves selected are best understood as the necessary consequences of other features of the organism. Panglossian just-so stories are inevitably likely to mislead.

The paper angered many of those present. It was attacked more for its irreverence and the presumed Marxist politics of its authors than for its content, although Arthur Cain, a long-standing student of snail evolution, responded by asserting that every one of the multifarious banding patterns observable on his snails must be adaptive; nothing was chance. Yet Gould and Lewontin's main point seems irrefutable – and until recently no one has even tried to rebut it. The exception is Dennett, who in his new book devotes the best part of a chapter (entitled 'The spandrel's thumb', in heavy parody of one of Gould's books, *The Panda's Thumb*)[30] to the attempt, arguing that, far from pendentives being necessary forms, there are a variety of possible space-filling designs which architects of cathedrals built with vaulted roofs could have employed; hence pendentives represent not inevitable but designed structures, generated by the architect for the religious purpose of depicting uplifting biblical scenes.

In arguing this way, Dennett labours but entirely misses the point. He might equally argue that there is nothing architecturally inevitable about the way the pendentives are decorated. That these paintings are 'adapted' to the religious needs of the community the cathedral served is obvious. However, Gould and Lewontin's case does not lie here, but in the argument that once the (architectural) decision has been made to mount a masonry dome on (orthogonal in the case of San Marco) arches, pendentives are inevitable. They are integral to the construction of arches and dome in a series of compressive rings, and *in situ* these curved, triangular elements take a substantial compressive force from the radial thrust of the dome.[31] The fact that the architect has some limited room for manoeuvre as to the exact form of the pendentives (such as merging with the dome or ending at a cornice) or could use a slightly different structural element (called a squinch) to bridge between arches and so provide a more nearly circular support for the dome is as precise an analogy as one might desire for both the strength and limits of adaptation. That is, adaptation is ultimately constrained by architecture, by limits imposed by forces outside the control of historical contingency. And it is to this even more fundamental critique of unrestrained adaptationist thinking that I now turn.

SELECTION IS NOT *À LA CARTE*

The arguments here flow, like many of the others in this chapter, from the descriptions of the lifelines in Chapter 6. Within the adaptationist programme, the trajectory that any lifeline can take is ultimately limited only by the question of whether it is adaptive. Of course, evolution is cumulative, and has to build on whatever materials it has to hand. So to arrive at any adaptive structure, behaviour or molecular property there has to be a legitimate route: from where the system is here and now, to some presumably more adapted place elsewhere. This route cannot run through a sort of adaptive valley between the present adaptive peak and the distant one – that is, the route between *here* and *there* must always be by way of forms at least as well adapted as those they succeed, or selection won't be able to work. The example is those word games in which one has to get, say, from CAT to DOG by changing one letter at a time in such a way that a valid word is produced at each step (e.g. CAT, COT, COG, DOG). This is why some of the structures one ends up with seem so cumbersome, and do not represent what would be recognized by engineers as 'good design'. The light-sensitive retina of the human eye is a good example. It is a seemingly back-to-front structure, and light only reaches it after having passed *through* layers of non-light-sensitive nerve cells, the results of both evolutionary and developmental history that would make any camera-designer wince. We carry the burdens of the past with us. None the less, granted the Dawkinsian assumption that 50 per cent of an eye is 1 per cent better than 49 per cent, adaptation will get there in the end. For ultra-Darwinists the menu of choice available to the adapting organism and species is essentially infinite.

The contrasting viewpoint is best expressed by Brian Goodwin and his long-time collaborator Gerry Webster. For them, profoundly under the influence of Waddington, evolution is uninterpretable except through the lens of *morphogenesis* – the development of the form of an organism. And morphogenesis is determined – or at least limited – by what they call, echoing a tradition in biology which predates Darwin, 'laws of form'.[32] At its most general, this argues that there are constraints deriving from principles of physics and chemistry on the possible degrees of freedom available to adaptationist selection. I have already described some of these constraints in action, in Chapter 6.

To take the simplest example, there is an ultimate limit on the size of any single-celled organism because of the physical fact that volume increases as the cube of the radius, whereas the surface area increases only as the square. All organisms need to trade with their external environment, for example by taking in foodstuffs and oxygen and excreting waste products and carbon dioxide, and this trading can be done only across the external cell membrane. As the volume of the cell increases, the problem of diffusing these waste products outward from the interior, and of the available surface area of cell membrane, becomes insuperable. The upper bounds of size for a single-celled organism are thus set by both chemistry and physics.

Similar constraints limit the size of multicellular terrestrial animals. The metabolic rates of organisms increase in proportion to body mass to the power $^{3}/_{4}$, and rates of heartbeat in proportion to body mass to the power $-^{1}/_{4}$. Times of blood circulation, embryonic growth and life-span vary as the $+^{1}/_{4}$ power of body mass.[32] Such general relationships are known as *allometric*. As their size increases, the dimensions of animals' bony skeletons must increase disproportionately in order to bear their weight without breaking – unless they take to the water to reduce the strain, which is why the largest animal that has ever lived, the blue whale, is indeed a marine and not a land organism. Reciprocally, as their size diminishes other constraints come into play, surface area grows large by comparison with body volume, and problems of energy conservation become serious. A humming bird's heart has to be relatively large in comparison with its overall body size, otherwise it would have to beat excessively fast.

Speed of movement, size, energy efficiency and hence behaviour are all shaped by physical constraints: for instance, an elephant has to sleep standing up because if it were to lie down its very weight would crush its own ribs. Depending on their size, different physical forces become more or less important. Watch a pond skater skimming the surface of a pool, and you are observing an organism whose very survival depends on the surface tension properties of the water supporting it (lower the surface tension by adding detergent, and the creature will sink). Gravity is relatively unimportant to the pond skater but very important to us humans, who can afford to be totally indifferent to surface tension. On a still smaller scale, single-celled organisms are buffeted by the Brownian motion of the molecules and ions in the fluid in which they are suspended, a type of force we can scarcely begin to comprehend from our own experience. On the other hand, neither pond skater nor single-celled organism is likely to be troubled by the effects of weightlessness during space travel; for them it would be pretty much business as usual.

It was J. B. S. Haldane who summarized differences most memorably in a famous essay entitled 'On being the right size', originally written in 1927, and from which I have drawn the epigraph for this chapter. Drop a horse, a man and a mouse down a mineshaft, and the horse disintegrates in mid-drop and, in Haldane's words, 'splashes' at the bottom of the shaft, the man is broken, and the mouse picks itself up and walks away. Such limits to the range of adaptation are not trivial. It is not merely original sin which prevents us humans from becoming angels. No range of musculature and load-bearing bones is possible which would enable organisms of our size and weight to sprout wings and fly. However hard it tried, evolution couldn't get us there.

While such constraints are pretty much self-evident, the idea of laws of form goes much deeper. An article in *New Scientist* in 1995 recounted with wonder how chemists had begun to synthesize crystal-like structures that closely resembled the marvellous delicate forms of some of the tiny *Radiolaria* species[34] (Figure 8.3). But such a resemblance is scarcely surprising. In his ground-breaking book *On Growth and Form*, first published in 1917,[35] the biologist D'Arcy Thompson first drew attention to the fact that radiolarian structures took these crystalline forms not, he argued, as the result of selection, but as a consequence of the workings-out of certain mathematically necessary constraints on crystalline growth.

Figure 8.3 Radiolaria as drawn by Haeckel. Note the regular, geometric forms of these tiny, quasi-crystalline organisms.

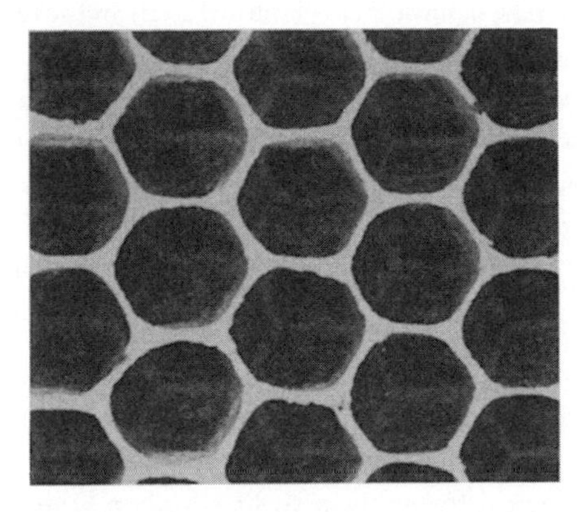

Figure 8.4 A honeycomb.

To see what these constraints might be, consider a simpler case, the honeycomb – a model example of a regular geometric structure – which was found in the eighteenth century, to the astonishment of those who studied it, to correspond precisely to half the form known to crystallographers as a rhomboidal dodecahedron (Figure 8.4), a so-called space-filling shape that makes possible the repeated close packing of the cells. How could the perfection of this structure be accounted for? For René Réaumur, in the 1750s, it was a clear example of planning and forethought by the bees that built it:[36]

Convinced that the bees use the pyramidal foundation which merits preference, I suspected that the reason, or one of the reasons, which made them decide in this way was to husband the wax; that among cells of the same size with a pyramidal base, the one that could be made with the greatest economy of matter or wax was that in which each rhomboid had two angles, each about 110° and two angles each about 70°.

It is the bees' knowledge of mathematics, according to Réaumur, that enables them to create these perfect structures. Today, ultra-Darwinists would be happy to rephrase this by postulating an adaptive gene for such rhomboidal construction. But wait. A mere 20 years after Réaumur, in the 1770s, the great biologist the Comte de Buffon was able to explain the phenomenon in different terms:[37]

Fill a vessel with peas, or any other cylindrical seed, and cover it closely after pouring in as much water as the spaces between the seeds allow; then boil the water; all the cylinders become six-sided columns. The reason, which is purely mechanical, is clear: each seed, which is cylindrical, tends to occupy as much space as possible in a given area; they therefore all become necessarily hexagonal by reciprocal compression. Each bee seeks to occupy the maximum space in a given area; since a bee's body is cylindrical it is necessary for the cells to become hexagonal for the same reason of opposing forces.

The point is that what seems to be an adaptation is in fact the inevitable result of physical forces which apply equally to inanimate and to living objects. D'Arcy Thompson generalizes the argument: the fact that many biological forms seem to fit simple mathematical or geometric rules indicates that the existence of constraining forces on their growth is a more parsimonious explanation than natural selection. Allometric formulae, which describe the ways in which different parts of an organism preserve their relationship to one another in related species differing in size, provide a good example. In Thompson's best-known examples, he shows that species of fish of very different apparent form can be shown to be structurally related to one another via rather straightforward topological transformations, as shown in Figure 8.5. The body plans of the fishes clearly fit their lifestyles,

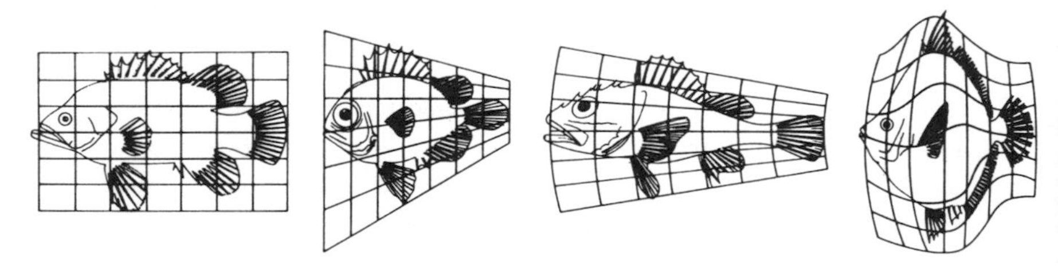

Figure 8.5 Topological transformations between four related fish.

but adaptive forces, in helping to generate them, have clearly been constrained by the availability of a finite range of topological transformations in creating a workable solution.

In other cases, the adaptive explanation clearly fails. What would an adaptationist make, for instance, of the fact that if you count the spiral rows of scales on a pine cone or seeds in a flower head you will find that they relate to one another according to the numbers of a famous mathematical expression, the Fibonacci series (named for the thirteenth-century Florentine mathematician who first defined it), in which each successive number is the sum of the two previous ones (thus 1, 1, 2, 3, 5, 8, 13, 21, . . .), a series which has also been employed in some striking modern art forms (Figure 8.6)? As Brian Goodwin points out, this is a pattern which can readily be generated within a relatively straightforward morphogenetic field.[38] Even if one could find an ingenious just-so story to account for the pattern, the sensible conclusion is that the adaptation is built around the structural constraint, and not vice versa.

Do such seemingly mathematical regularities account for other characteristic aspects of living morphologies? Goodwin and Webster have argued that they do, citing for example the characteristic tetrapod limb form shared by all vertebrates. One distinctive feature of such

(a) (b)

Figure 8.6 The regular array of components spiralling in the Fibonacci series. (a) A pine cone. (b) A shasta daisy flower head.

limbs is that they all start at the shoulder or hip with a single bone, respectively the humerus or femur. No known fossil or living vertebrate ever had two, although such a structure would presumably be very useful to birds which need a flat, light, strong structure for their wings, and two struts can be lighter and stronger than one. Models of the process of making a tetrapod limb suggest why they always start with a single bone. Goodwin argues that this is an example of such a 'law of form' in action, yielding the characteristic tetrapod limb as a stably generated structure with which only minor adaptive tinkering is feasible. If the adaptationists see but one pattern in the kaleidoscopic variety of living forms, Goodwin shakes the kaleidoscope to reveal another, perhaps no less plausible.

Challengingly, and to me quite unacceptably, Webster has declared that the ultimate goal of this approach to the problem of biological form is to eliminate the historical accounts provided by evolution entirely, replacing them with such 'laws of form'. Selection, far from being *à la carte*, is then limited to the narrow choice of menus offered by a mathematical *table d'hôte*. Evolutionary biology will then become, in his phrase, 'mere antiquarianism', a trivial picking over the residues from the combination meals chosen by past diners at the feast of life. I remain to be convinced. I would still insist on my modified version of Dobzhansky: 'Nothing in biology makes sense except in the light of history.' Ah, yes, but a history far richer than is offered by mere adaptationism.

ORGANISMS AS ACTIVE PLAYERS IN THEIR OWN DESTINY

When Karl Popper incurred the wrath of the evolutionary biologists assembled at the Royal Society meeting that I described in Chapter 4, he did so by counterposing what he called 'active' and 'passive' Darwinism. By this I understood him to mean that he saw organisms as doing more than merely responding passively to environmental pressures – and if that is not what he meant, it is certainly what I mean. In ultra-Darwinism, organisms have an inherent passivity, helplessly ground between the nether and upper millstones of their genetic endowment on the one hand, and subject to the impersonal winnowing force of natural selection on the other. The entire metaphor of natural selection is one in which 'Nature' (a.k.a. God) sets a series of challenges which organisms either meet, in which case they are privileged enough to pass on copies of their genes to a successor generation, or to which they succumb, in which case they leave merely their material bodies to be recycled and provide challenges and resources for other, scavenger organisms. As Darwin put it, nature is constantly subjecting living forms to ruthless 'scrutiny'.

By contrast, the picture I have tried to paint here, building on the autopoietic description of lifelines in Chapter 6, is one in which organisms do not sit waiting patiently for nature or 'the environment' to scrutinize them, but rather are actively engaged in working to choose and transform their environments, to adjust and appropriate them to their own ends. Autopoiesis, organisms as active players, is as apparent when a single-celled organism swims away from a depleted food source towards a rich one as it is when a growing troop of axons from the retina of a cat seek, find and modify their target neurons in the lateral geniculate, in the symbiotic relationship of a leguminous plant with the nodules of nitrogen fixing bacteria in its roots, and in the decision of an impoverished Mexican to cross the border into California or an unemployed Newcastle builder to move to Düsseldorf.[39] This is not passive acceptance of anything the Great Selector throws their way, but an essential aspect of their nature as living organisms. Nor is it of course a statement about purposive and conscious attempts to direct evolutionary processes; I am not resurrecting Teilhard de Chardin or anthropic principles, and if that was what Popper meant by active Darwinism then it couldn't be further from my intentions. I am, however, asserting the part that individual organisms play in shaping their own future: how it is that, if biology is indeed destiny, then that destiny is constrained freedom.

It should also be apparent by now that 'environments' are not static and unchanging, but are themselves undergoing constant change as a very result of the work done on them by living processes. This is why Dawkins, with his gene's-eye view of the world, is able to describe environments as the extended phenotypes of the organisms that inhabit them. In some ways this is not a bad concept – provided one recognizes that it carries within it the seeds of destruction of the individualistic gene's-eye view, for such an environmental phenotype is by definition the shared phenotype of many genotypes. Nothing could be further from the truth than the picture often painted by environmentalists of a natural world which, were it not for human intervention, would persist in a condition of harmonious stasis, unchangeably 'in balance'. Homeostasis – the 'balance of nature' – is as misleading a metaphor for environments as it is for organisms: homeodynamics is the order of existence. Environments have their own trajectories – lifelines, if one is an enthusiast for James Lovelock's Gaia metaphor – constantly being transformed not merely by the workings-out of the inanimate forces of weather, temperature and cosmic history, but above all by the interactions of myriad life forms.

BEYOND ULTRA-DARWINISM

To summarize: The metaphysic of ultra-Darwinism rests on premises which combine a theology of preformationism with a belief in the invisible hand of the market *à la* Adam Smith to produce a Panglossian vision in which a competitive struggle of all against all at the level of individual genes produces the rich diversity and relative homeodynamic tranquillity of a living world which is nothing more than the extended phenotype of these selfish genes. In contrast, I have argued that:

1. The individual gene is not the only level at which selection occurs.
2. Natural selection is not the only force driving evolutionary change.
3. Organisms are not indefinitely flexible to change; selection is at least in some measure '*table d'hôte*' and not '*à la carte*'.
4. Organisms are not mere passive responders to selective forces but active players in their own destiny.

In the next chapter I consider how these alternative views of living processes affect our understanding of our own modern origin myths. What is life and how did it originate on Earth?

NOTES

1. Robert Wright, '20th century blues'.
2. Edward O. Wilson, *Sociobiology: The New Synthesis*, p. 553.
3. Ferren MacIntyre and Kenneth W. Estep, 'Sperm competition and the persistence of genes for male homosexuality'.
4. Richard Dawkins, *The Selfish Gene*, p. 215.
5. Wilson, *On Human Nature*, p. 172.
6. Brian Goodwin, *How the Leopard Changed Its Spots*.
7. Richard Lewontin and Richard Levins, 'The problem of Lysenkoism'.
8. Ernst Mayr is Dobzhansky's philosophical and scientific heir, and his insistence on a consideration of the genome as a proper level for selection characterizes much of his writing. See e.g. his *Towards a New Philosophy of Biology*.
9. James A. Shapiro, 'Adaptive mutation: Who's really who in the garden?'.
10. Conrad H. Waddington (ed.), *Towards a Theoretical Biology*.
11. John Tyler Bonner, *On Development*.
12. I. Wilmut *et al.*, 'Viable offspring . . .'.
13. Lancelot Law Whyte, *Internal Factors in Evolution*.
14. Bonner, *The Evolution of Complexity by Means of Natural Selection*, p. 93.
15. Studied by a statistical technique known as quantitative trait locus analysis – see S. D. Tanksley, 'Mapping polygenes'.
16. J. L. Hubby and R. C. Lewontin, 'A molecular approach to the study of genic heterozygosity . . .'.
17. Stephen Jay Gould and Niles Eldredge, 'Punctuated equilibria'.
18. There have been very recent claims to have demonstrated punctuated equilibrium occurring in test-tube models of bacterial evolution. See Santiago F. Elena, Vaughn S. Cooper and Richard E. Lenski, 'Punctuated evolution caused by selection of rare beneficial mutations'.
19. The reasons are summarized accessibly, though excessively sociobiologically, by Matt Ridley in *The Red Queen*.
20. John Maynard Smith, *Did Darwin Get it Right?*, Chapter 22.
21. Mayr, *The Growth of Biological Thought*.
22. Virginia Morell, 'Genes *v.* teams: Weighing group tactics in evolution'.
23. James H. Tumlinson, W. Joe Lewis and Louise E. M. Vet, 'How parasitic wasps find their hosts'.
24. N. Moore, *The Bird of Time*, pp. 124–5.

25. A term she resurrected from the Russian literature, which she arranged to have translated: L. N. Khakina, *Concepts of Symbiogenesis.*
26. At the DNA level, the geneticist Gabriel Dover has been arguing for several years now for a process he calls molecular drive. Based essentially on the molecular properties of DNA, the drive argument claims that, over time and replications, changes will occur in DNA sequences which are independent of selective forces operating externally to the molecule. The result is that drift occurs not merely at the level of the phenotype but at the genomic level too. The explanation for this effect is more technical than can easily be encompassed within my description here, and it is fair to say that it has not yet achieved wide support among molecular biologists; I include them here for the sake of completeness in this account.
27. Gould, *Wonderful Life*, p. 14.
28. The debate between the two views is analysed in Richard C. Lewontin, *The Genetic Basis of Evolutionary Change.*
29. Stephen Jay Gould and Richard C. Lewontin, 'The spandrels of San Marco and the Panglossian paradigm'; the responses I refer to are not part of the published proceedings.
30. Daniel C. Dennett, *Darwin's Dangerous Idea*, Chapter 10.
31. I have been troubled enough by Dennett's argument to consult a number of architects on this point, and their response has been unanimously in favour of the original Gould–Lewontin interpretation. I am particularly grateful to Renate Prince for researching this question more deeply than I have space to do justice to here, and for a most instructive tutorial, and also for the relevant reference: Rowland Mainstone, *Developments in Structural Form.*
32. Goodwin, *How the Leopard Changed Its Spots.*
33. Geoffrey B. West *et al.*, 'A general model . . .'.
34. Philip Ball, 'Spheres of influence'.
35. D'Arcy W. Thompson, *On Growth and Form.*
36. The quotation is as cited by François Jacob in *The Logic of Living Systems.*
37. Comte de Buffon, *De la manière d'étudier et de traiter l'histoire naturelle*; cited in this translation by Jacob, *The Logic of Living Systems.* It has been pointed out to me by an anonymous reviewer that Buffon oversimplifies here. Bees and wasps use different methods to form hexagons, and neither uses its body in the way that Buffon implies. But Buffon's point – and D'Arcy Thompson's – remains: a space-filling model of the sort that the hexagonal forms generate can be derived simply from a consideration of the physical constraints involved without the need to invoke adaptive mechanisms whereby previous generations of bees constructed a variety of inferior, less

efficient forms which evolved by natural selection to produce the current state of 'perfection'.

38. Goodwin, *How the Leopard Changed Its Spots*, Chapter 5.

39. This phrase has been criticized by some reviewers, rightly, I think, for implying homology of process. Of course I had not meant to reduce the personal and social processes of human choice in all their rich complexity, to merely the motion of unicells!

9

Origin Myths

Which came first, the chicken or the egg?
Traditional riddle

CHICKENS AND EGGS

In some ways, this book has all been about chickens and eggs: chickens as the egg's way of making another egg, or eggs as the chicken's way of making another chicken. Ultra-Darwinists are unequivocal – primacy goes to the egg. Much of the argument presented in the previous chapters of this book has served to restate the chicken's case against what appears to be the dominant grain of current biological thinking.

Speculation about the origin of life of course goes back far beyond present-day biology. It forms part of the creation myths of most cultures: the first humans, for instance, fashioned on a potter's wheel from mud or clay, into which a creator-god breathes the breath of life. Until the last couple of decades, there was a strange continuity between such myths and biology's origin stories. The definition of being alive was to be a breathing, metabolizing, environment-sensing and responding organism. However, most modern molecular biologists will have no truck with such ideas. For them, the basic function of life is narrowly defined as the power to replicate, and the basic unit of life is therefore a molecule with this power, a naked nucleic acid polymer. Granted that the significance of a replicator may be narrowly defined in terms of the message conveyed by the string of letters signifying the nucleotide bases, a certain theological pricking of the ears may occur at this point. What replicator theory is telling us, quite unabashedly, is that, in the phrase of the Gospel according to John:[1]

> In the beginning was the Word, and the Word was with God and the Word was God. The same was in the beginning with God. All things were made by him; and without him was not anything made that was made. In him was life . . .

For the three letters of GOD, substitute DNA's four: ACGT. In the Jewish religion within which I was raised, it was sacrilege to speak the hidden name of God except on the sacred occasion of the Day of Atonement, Yom Kippur. Today's molecular biologists, however, with all their Frankensteinian insouciance, have no qualms about not merely speaking but even manipulating the sacred letters; no longer the mud out of which the potter fashions life, they have taken upon themselves the responsibilities of the potter. Despite this subliminal quasi-theology, the naked replicator view of life's origins has the imprimatur of such distinguished molecular biologists as Francis Crick and Leslie Orgel, quite apart from the philosophers and popularists who trail in their wake, and it would seem to require a certain amount of pig-headedness to oppose it.

NUCLEIC REPLICATORS

Of course, the problem is partly semantic. If you define the basic property of living systems as the capacity to reproduce exact equivalents of themselves, then attention is automatically directed towards those molecular or supramolecular structures which are capable of achieving such precise copies. The definition inevitably centres the debate around the origin of the nucleic acids because, as far as is known, of all the molecular and macromolecular species present in currently living organisms, only they possess the potential for such replication. Until a few years ago, the nucleic acid concerned was believed to be DNA; today there has emerged a powerful counter-school of speculation which argues the case for RNA. An 'RNA-world' is held to have preceded today's DNA world for reasons which make a certain biochemical sense.

As I have pointed out, in themselves DNA and RNA are both stable, inert molecules. To make copies of DNA requires not merely the DNA molecule but an array of enzymes, brought together in close proximity and within a rather closely controlled environment. The same is in principle true for RNA, but unlike DNA it is single-stranded, rather than a double helix, and so perhaps it is easier to envisage it being synthesized by relatively unsophisticated systems. Furthermore, the discovery (referred to in Chapter 3) that some forms of RNA – ribozymes – can function as enzymes raises the intriguing possibility of the first 'living' replicator being an RNA molecule which possessed the enzymic power to catalyse the synthesis of copies of itself – an auto-ribozyme, one might say. Once this self-copying power had developed, the auto-generative engine of natural selection would inevitably be brought into play, ensuring that those auto-ribozymes which could copy themselves most rapidly and most faithfully, under the prevailing environmental conditions, would survive and multiply. And the rest, in this scenario, would be history.

Could such a system work? Could life have begun with an auto-ribozyme which could haul itself up by its own bootstraps? Well, test-tube experiments have shown that artificial selection can result in the evolution of RNA sequences. Take an appropriate mix of precursors, 'primer' RNA sequences and enzymes, including the vital RNA polymerase, and allow RNA synthesis to proceed. After a time, stop the reaction, isolate the RNAs and select only those of a specific chain length, and set up the next test-tube with these in place. Repeat the selection procedure a few times, and you will end up with an RNA synthesizing mix which preferentially produces RNAs of the chain length that you have arbitrarily set and selected for. Despite the intriguing nature of these experiments, however, they don't really answer the question of origins, any more than the artificial selection methods used by plant and animal breeders resolve the problem of natural selection. The biochemical systems that catalyse such evolving RNA syntheses are already quite complex. They must occur in a test-tube, which serves as a surrogate cell, including the necessary mix of enzymes, ions and controlled temperatures. The nearest one gets to naked replicators today are the DNA or RNA viruses, and, as is clear, these too can sit indefinitely in test-tubes as crystalline powders without ever being able to replicate. Purity in nakedness is sterile.

It follows that accurate replication could not have emerged until long after the development of cell-like structures capable of such crucial living processes as metabolism, growth and division. The Earth is said to be some 4.5 billion years old. The earliest cell-like structures that have been found can be dated to about 3.5 billion years ago, a mere 300 million years after the Earth's crust had cooled to below the

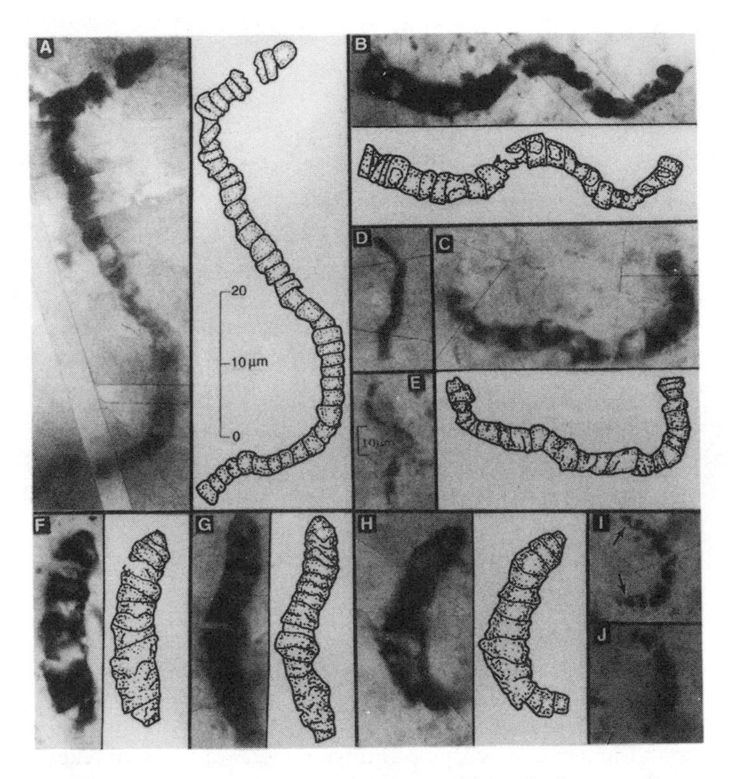

Figure 9.1 Fossil cells, 3.4 billion years old. Each photograph is accompanied by a diagrammatic reconstruction.

boiling point of water, and under the microscope they look pretty similar to some of today's bacteria (Figure 9.1).[2] Although there is as yet no way of knowing whether these ancient cells contained nucleic acid replicators, what characterizes them above all is the presence of a cell membrane that provides a boundary between the interior and exterior of the cell. It is interesting that it is the interpretation of the tiny structures found in a Martian meteorite as possessing such boundaries in fossilized form that has led NASA scientists to interpret them, however dubiously, as indicating the presence of life on Mars (Figure 9.2).[3] I shall argue that it is the presence of this cell membrane

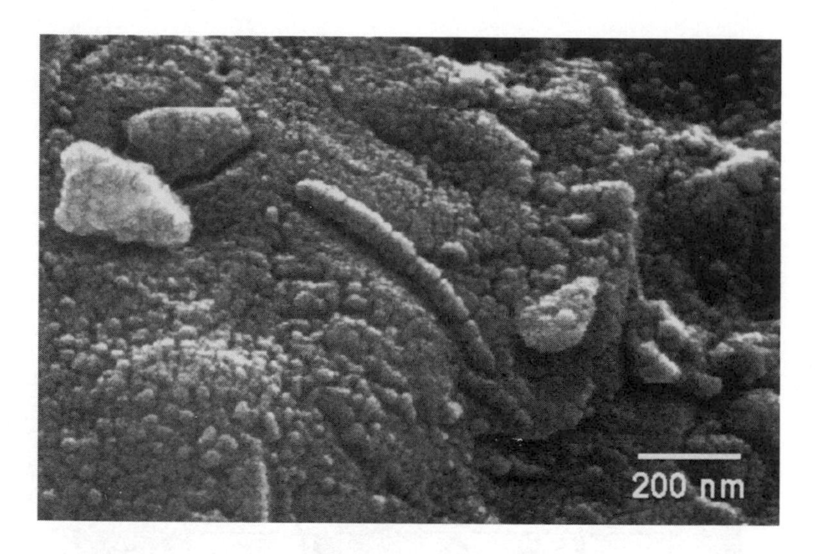

Figure 9.2 Structures in a meteorite from Mars, claimed to be evidence of fossilized primitive life (1nm = 10^{-9} metres).

boundary, rather than replication, which must have been a first, crucial step in the development of life from non-life, for it is this that enables a critical mass of organic constituents to be assembled, making possible the establishment of an enzyme-catalysed metabolic web of reactions. Only subsequently could accurate replication based on nucleic acids have developed.

CHEMICAL CHANCE OR NECESSITY?

So how might one get from a distinctly non-living, slowly cooling Earth to the origin of cells? The first attempts to think this problem through systematically date from the 1920s, and were by the biochemist Alexander Oparin,[4] in what was then the Soviet Union, and that Renaissance man of British biology, J. B. S. Haldane.[5] One of the key features of living systems which any theory of origins is required to explain is how it is that, of all the vast numbers of possible organic molecules and reactions which might conceivably characterize them, only a tiny fraction take part in the biochemistry of all the diverse species which have hitherto been studied. The utilization of sugars, especially glucose, as a principal energy source, the sequence of reactions through which it is transformed, and the synthesis of ATP as an immediate energy source are almost universal. Only occasionally are sugars other than glucose, or 'energy-rich' compounds other than ATP, found playing a major metabolic role. Of all the many amino acids, only about twenty are naturally occurring and serve as the building blocks for proteins. Furthermore, both sugars and amino acid molecules can each exist in two almost identical forms – optical isomers – known as D- and L-forms, a terminology based originally on an observation by Louis Pasteur concerning the direction in which pure crystals of the isomers rotated the plane of polarized light. But naturally occurring sugars are all in the D-form, while the amino acids are all in the L-form. How is one to account for these exclusions?

When one turns to the macromolecules, the mean-spiritedness of biochemical nature is even more striking. The numbers of potential proteins which could be assembled from these twenty amino acids is so vast as to beggar comprehension. A modest protein, with a molecular weight of some 34,000 and containing combinations of only 12 of the naturally occurring amino acids, could exist in 10^{300} possible forms, and if only one molecule of each existed the total mass would be around 10^{280} grams – compare this with the mass of the entire universe, which one estimate puts at 'only' 10^{55} grams! The actual is vastly outnumbered by the potential. And yet, with all this range of potential open to them, the numbers of different proteins found in organisms as diverse as bacteria and whales is really quite small. While no one knows the actual number, since only a fraction of species have ever been studied biochemically, it would be surprising if it amounted to more than a few tens or hundreds of millions of generic forms at the outside. Of the hundred thousand or so different proteins in humans, for instance, most can be found in virtually all other animals studied, with subtle variations depending on the speed of the molecular mutation clock and on the length of time since the evolutionary divergence of the species from a common ancestor.

There are many quite deep implications of this surprising observation. Anti-evolutionists, like the cosmologist Fred Hoyle and US creationists, have used it to argue that life cannot possibly have arisen by random purely physico-chemical processes. Hoyle has likened the chance of synthesizing a specific protein in this way as equivalent to that of a hurricane assembling a jumbo jet from its components laid out in an aircraft hangar.[6] There simply hasn't been enough time since the formation of the Earth for such processes to have come about as the result of random syntheses. Hoyle and others have therefore been attracted to the idea that our planet was 'seeded' with already living forms from space, perhaps delivered by meteorite or comet – the so-called 'panspermia' hypothesis. I've always found this proposition silly, even when it has come from the pen of Francis Crick.[7] For it doesn't solve anything to push back the question of origins by however many billions of years the panspermia idea might buy for evolution. The odds against chance assembly, even elsewhere in the universe, would still be far too great. The argument is *in principle* specious, just as is the 'half an eye' case described by Dawkins. It is as if, having abandoned the argument from physiological or anatomical design offered by Paley, these modern anti-evolutionists instead take refuge in biochemical complexity, the exquisite coordination of metabolic pathways and enzyme interactions.[8] Transferring the problem from anatomy to biochemistry, however, does nothing fundamentally to alter the nature of the issues at stake, which are surely simply that evolutionary processes are not *à la carte*, but constrained by chemical and physical properties.

To begin with, there are almost certainly chemical constraints on the range of available building blocks for living processes which are every bit as significant as the structural constraints discussed in the last chapter, but of which science is at present almost entirely ignorant. Thus it may be that the naturally occurring amino acids are simply those which are most readily synthesized abiotically, or their close relatives. Their existence in only one of their possible optically isomeric forms does require some ingenious chemical explanation, it is true, as most abiotic syntheses produce the two forms in equal proportion. However, what is clear is that once one of the two possible forms has emerged, it would rapidly have had to become universal. Because of the close molecular resemblance of the optical isomers, the 'unnatural' varieties can easily bind to the active centres of enzymes which normally catalyse the conversion of the naturally occurring forms. Once bound, however, they clog up the enzymes' active sites, and thus act as metabolic poisons. In a world in which all organisms depend for their survival largely on their biochemical compatibility with one another, organisms either depending on or producing the 'unnatural' forms, except as very specific toxins, would soon die out.

As for proteins, the ordering of amino acids within them is not random assembly in any old order by courtesy of a hurricane. Certain sequences are preferred in that they fall into appropriate three-dimensional configurations, and can self-assemble as discussed in Chapter 6. In most protein molecules there are parts which seem to be highly conserved in evolutionary terms, suggesting either that they conform to least-energy configurations, in which case their 'selection' during evolution depends on physico-chemical constraints of the sort discussed in earlier chapters, and/or that they are essential to the enzymic or structural function of the protein, such that mutations would be deleterious or even lethal. An example is the valine–glutamate substitution in sickle-cell anaemia, which I remarked upon in Chapter 2. But other regions of the peptide sequence of proteins are quite highly variable, both within and between organisms of the same species – the isoforms already referred to – or between species. This suggests that, like intron DNA, they may be contingent accidents of history, functionally without great significance, and therefore not requiring the engineering precision on which Hoyle's analogy depends.

The implication of biochemical parsimony, of the limits of the actual compared to the potential which is of significance to my thesis here, is very different. It is that, whatever the known or guessed-at physico-chemical constraints which provided the parameters within which primeval molecules originated, much of the biochemical evolution to which we humans and all other life on Earth are today's heirs must have occurred *before* our distant evolutionary ancestors separated into the great morphologically distinct kingdoms. The extent to which we share a common biochemistry with oak trees, bacteria and yeast cells reflects genuine homology – a common ancestry. Hence whatever organic chemicals the primitive biosphere contained, there must at some point before the great divergence have been an evolutionary bottleneck which excluded all but a small subset of the potential chemical building blocks of cellular life.

If the range of organic chemistry with which biologists have to deal is thus extraordinarily parsimonious, the limitation of the numbers of inorganic chemicals used is perhaps even more striking. Life consists primarily of arrangements of the elements carbon, hydrogen, oxygen and nitrogen, together with smaller quantities of phosphorus and sulphur, and ions of calcium, magnesium, sodium, potassium and some heavy metals such as iron, zinc and copper – some fifteen elements in all. The chemist R. J. P. Williams has recently considered the evolutionary implications of these limitations, basing his case on the observed abundance of the elements on Earth and their chemical properties, in a ground-breaking book provocatively entitled *The Natural Selection of the Chemical Elements*.[9] Thus he points out that not merely are hydrogen, carbon, nitrogen and oxygen both abundant and available, but the compounds they form have specific properties relevant to life. In particular, provided energy sources are available, they combine readily to form thermodynamically unstable compounds, capable of relatively long life in watery solution; they trap usable energy easily in the form of sugars; they readily combine to form long chain molecules – lipids, polysaccharides, proteins and nucleic acids. Incorporation of phosphorus and sulphur dramatically extends the range of available compounds and the richness of their interactions, as does the addition of metal ions, and Williams proposes an evolutionary sequence of increasingly complex chemical interactions available to and sequestered within proto-cells. For him, as for Oparin and Haldane, the question of origins lies not with the development of replicators, but with the development of the cells that contain them and with their chemistry.

Even inorganic materials are, it appears, capable of synthetic reactions resulting in complex forms. It has been suggested that concepts such as morphogenesis, replication, self-organization and metamorphosis can be applied to such chemical syntheses, based on micelles, vesicles and foams, as Figure 9.3 shows.[10]

ABIOTIC SYNTHESES

Biochemical knowledge was of course far less sophisticated when Oparin and Haldane advanced their theses. But the problems they faced were genuine enough. First, if life was to evolve from non-life, the conditions for abiotic synthesis of organic compounds which might subsequently serve as the basis for cellular development had to be present. But this assertion reawakened the metaphysical dispute which biologists had reckoned was settled when, more than half a century previously, Louis Pasteur had convincingly refuted claims for spontaneous generation. *Ex ovo omnia*, William Harvey (discoverer, for Western medicine, of the circulation of the blood) had affirmed three centuries before, but it took Pasteur to clinch the argument by showing that, in sterile conditions and with contaminants rigorously excluded, the conditions that otherwise favoured fermentation and the appearance of bacteria and moulds now failed to do so.[11] But if all life comes from life, how – except by divine intervention – could it have arisen in the first place?

Darwin himself had recognized the need to circumvent this paradox when he speculated that, even as he was writing, the abiotic precursors of life's chemicals might be being synthesized in some small warm pool, the drying margin of the sea. What Darwin did not realize, not

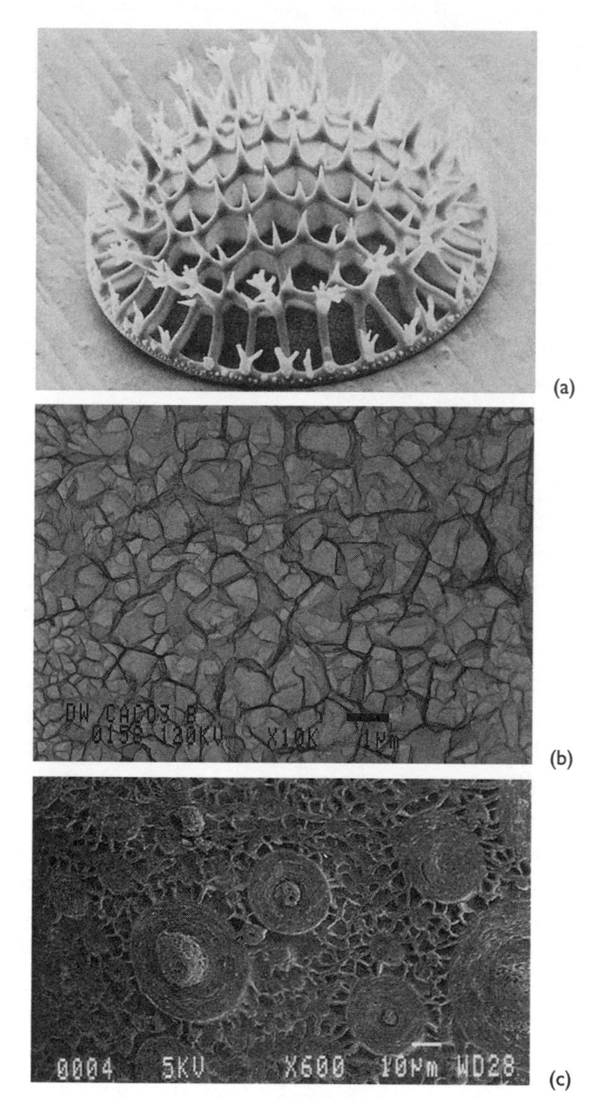

Figure 9.3 (a) The valve of a diatom. (b) A calcium carbonate membrane formed on an oil/water foam. (c) Aluminophosphate vesicles forming complex synthetic patterns.

being a chemist, was that in an Earth whose atmosphere contains so much oxygen such syntheses would be almost impossible to achieve. Oxidizing environments are pretty toxic, except for those life forms like ourselves which have evolved the special facility to exist in them. What is required is an atmosphere with something like the composition that the Galileo probe has observed on Jupiter – a reducing mixture of hydrogen, ammonia and methane – together with carbon dioxide. And that, Oparin argued, was precisely what the Earth's primitive atmosphere was like. The present-day atmosphere has replaced the primitive one precisely because of the action of life itself, the photosynthesizing work of plants over thousands of millions of years. This insight, of the power of life to change the chemical and physical composition of the Earth itself, long predated James Lovelock's Gaia metaphor, developed during the 1960s and 1970s,[12] although of course it was developed within a very different metaphysic.

Modern-day plants take up carbon dioxide and use it to provide the carbon skeletons of the sugars and lipids they require, and in the process release oxygen. The nitrogen source for their proteins and nucleic acid is first 'fixed' from atmospheric nitrogen as ammonium salts by plants and bacteria. In a carbon dioxide, methane and ammonia-rich atmosphere such syntheses would have been relatively easy, but as the mass of 'fixed' organic compounds increased, so the composition of the atmosphere would have begun to change, gradually becoming rich in oxygen and poor in carbon dioxide. This process has been continuing steadily over the last three and a half billion years of life on Earth (and is now being partially reversed by the results of industrialization and the release of 'greenhouse gases', notably carbon dioxide, into the Earth's atmosphere). In discussing this process, Margulis has analogized it to the autopoietic, self-constructing capacity of individual organisms, and described it as ecopoiesis.[13] Just as organisms construct themselves and species evolve, so too do environments, regulated homeodynamically. Only after photosynthesizing organisms had changed the Earth's atmosphere would oxygen-requiring life forms have evolved, to live, as we and all other animals do, off the back of the photosynthesizing work of the plant kingdom. But the essential point is that the biochemical versatility available to all current living forms – including their capacity to reproduce – could evolve only on the basis of already well-established and sophisticated metabolic webs and enzymes.

In the warm oceans beneath the reducing atmosphere of the primitive Earth, bombarded by violent electrical storms, Oparin argued, a multitude of organic chemicals would begin to be synthesized, although they would remain distributed in weak solution. It was not until many years later, in California in the 1950s, that this theoretical observation was given some experimental substance. In a famous experiment, Stanley Miller[14] placed a gas mixture of hydrogen, methane and ammonia in a sealed flask, warmed it, and passed an electrical discharge through it for twenty hours or more, thus attempting to mimic primeval lightning. At the end of this time his flask contained amino acids and other organic acids, the potential building blocks of life. Later modellers of abiotic syntheses, such as Sydney Fox, opted for alternative routes such as dry synthesis – reproducing the conditions that would have obtained had the chemical mix been fired in the heat of a volcano. Under such circumstances too, they found that tarry messes containing both simple organic compounds and even some peptides and ATP could result. Rainstorms would dissolve these dry synthesized chemicals, and wash them into pools. 'Origin of life' conferences have been enlivened by debates between protagonists of these alternative routes. As I am not sure that I see how the issue can ever be resolved, I regard the dispute as taking on something of the form of the argument between Swift's Lilliputian big- and little-endians concerning the correct entry point into a boiled egg. In any event, for my purposes here it doesn't matter which, or if both, are correct. The point is simply that such abiotic syntheses of the basic chemical constituents of living forms can occur, and that plausible materialist accounts of how life could have originated on Earth can therefore be provided.

However, neither route would provide precursors for life unless the weak solutions could be concentrated in some way. The drying margins of seas might be a possibility, and the surfaces of rocks and clays containing metal ions might serve as the catalytic surfaces on which compounds could be concentrated and their metabolic transformations could begin. The chemist A. G. Cairns-Smith has built a convincing case based on the known chemistry of clays to show these might have provided the crucial surfaces and catalytic powers required.[15] Perhaps the metaphor of the mud, the potter and the wheel might not be such a bad one for creation after all.

COACERVATES

But how to get from here to cells? This is where Oparin's second insight became important. The chemistry and physics of large molecules like lipids and proteins were not well understood when he wrote; instead, their strange properties in solution were studied under the name of colloid chemistry. It was known that solutions containing such large molecules have a remarkable tendency to break up into small droplets containing the polymers in concentrated form, leaving the surrounding medium comparatively free of dissolved substances. Salts and organic molecules of low molecular weight in the solution also tend to get sucked into these droplets. Perhaps the droplets could even take up clay granules, with their catalytic surfaces. The phenomenon, of much less interest now to chemists than it was in the early part of this century, is called *coacervation*, and its products are known as *coacervate drops*.

Oparin argued that just such coacervate drops would begin to be formed from the dilute solutions containing polymeric organic compounds. The organic material present would be concentrated within them, enabling a critical mass to be achieved, and thus metabolic interactions between the compounds could begin. Some of these droplets would become unstable as a result of the reactions and would break up; others would be more stable and would continue to attract material until they exceeded some critical size, whereupon they would split into two daughter droplets each containing something approximating to the mix in their parent. Replication without a naked replicator would have been achieved.

Coacervate drops and colloid chemistry are distinctly out of fashion today, despite the attractions of Oparin's mechanism. And his droplets are still without an external bounding membrane, which I have argued is the *sine qua non* of cellular life. But such membranes can be created abiotically – and indeed are so created every time a drop of oil or lipid is placed in water. Depending on the amount of oil relative to the water, it either forms a thin film on the surface, or coalesces into a small droplet in which the lipid molecules align themselves precisely as they do in the external membrane of a cell. This property, of creating so-called liposomes, is today exploited to encase the naked strands of DNA intended to be inserted into cells during genetic engineering experiments (for instance in the treatment of the disease cystic fibrosis, the result of a deficient gene). The liposomes containing the hoped-for remedial genes fuse with the external cell membrane of their target cells and release their contents into the interior. The cell membranes apparent in the microscope pictures of the bacteria of 3.5 billion years ago (Figure 9.1) could have been generated in precisely this way.

Thus coacervation would concentrate inorganic ions, organic chemicals and simple polymers out of the dilute seawater solution, and liposome membranes would form around them. The first proto-cells

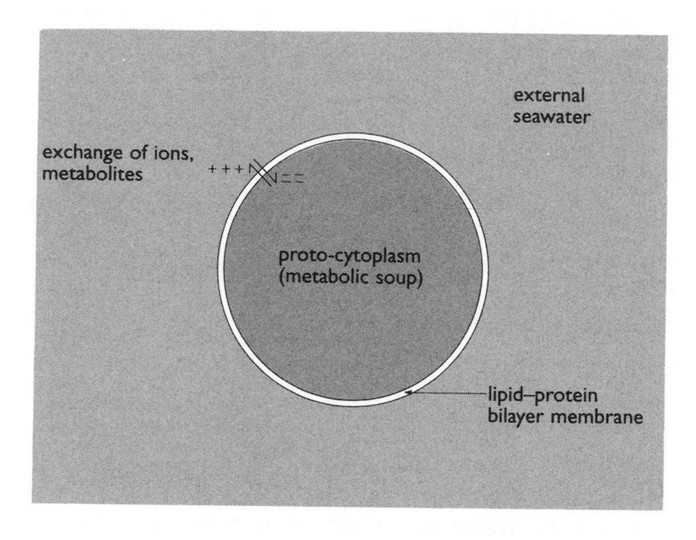

Figure 9.4 A possible proto-cell.

would have emerged. Such cells would have another property seemingly fundamental to life. The distribution of electrically charged ions, such as the positively charged sodium, potassium and calcium, and negatively charged chloride (all present in seawater) across their membranes would be asymmetric for basic physico-chemical reasons (Figure 9.4). This asymmetry ensures the seemingly universal property of living cells, of being some 65–95 millivolts negative to the outside. The significance of this electrochemical gradient in helping to concentrate certain substances within the cell, and excluding others, cannot be overestimated.

CATALYTIC WEBS

The next evolutionary step would be to stabilize the myriad potential chemical reactions that could occur within the proto-cells. This process has recently been modelled by Stuart Kauffman.[16] He makes a number of plausible assumptions about the behaviour of such a chemical soup. For instance, given a sufficient number of different compounds concentrated within a lipid membrane, even without the potential enhancement of catalytic surfaces such as those provided by inorganic substances like the clays, a tiny number of the molecules present will be able to function as catalysts for reactions between other components. In some cases, there would arise autocatalysis, in which a substance catalyses its own synthesis, or mutual catalysis, in which one substance catalyses the synthesis of another, which in its turn catalyses the synthesis of the first. Indeed, certain peptides have been found to show just such autocatalytic properties.[17] Computer models of such processes show that, with these catalytic assumptions, a random set of chemicals in a constrained area soon settle into a robust and autopoietic metabolic web, of the type described in Chapter 6, in which stable balances of constituents result (Figure 9.5). The consequence is homeostasis – a necessary precondition for homeodynamics. Traffic across the liposome membrane will bring new materials into the cell and excrete waste products, and, just as with the coacervate drops, cells which increase in size will simply split into two.

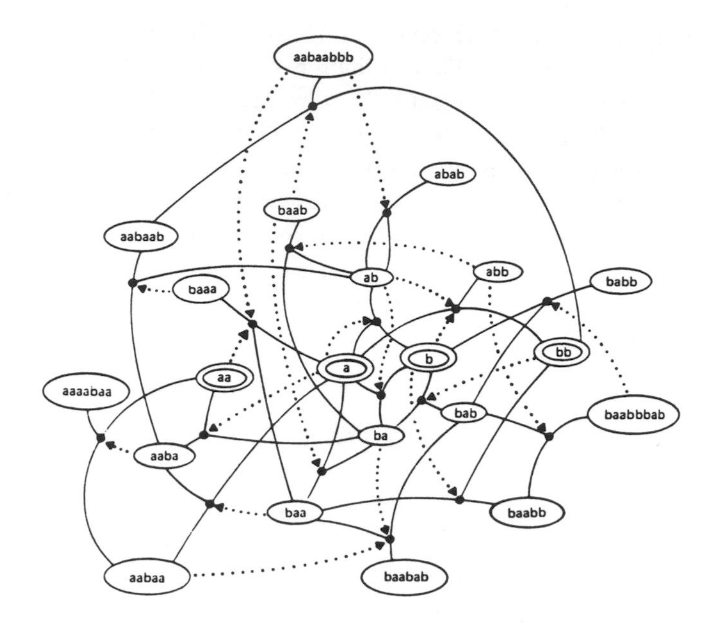

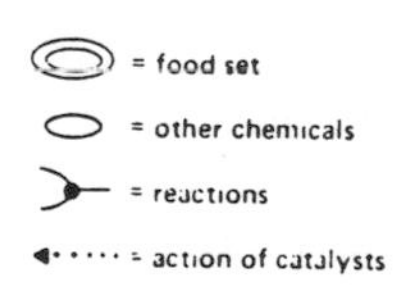

Figure 9.5 An autocatalytic set. Food molecules (a, b, aa, bb) are built into a self-sustaining network. Reactions are represented by points connecting large molecules to their breakdown products; dotted lines indicate catalysis.

So far we have done without molecular replicators altogether. Cell formation and division, and indeed sophisticated metabolic stability, have all been achieved by originally abiotic processes in which the properties that characterize life are captured not in a single molecule, but in the entire system which constitutes the cell. Indeed, one can go further. The metabolic web must have extended beyond any individual proto-cell, to embrace the entire living population of proto-cells. For just the same type of reason that once particular optical isomers of amino acids and sugars had emerged, they had to become the standard form, so too must the vast bulk of the constituents of the metabolic web. For if chemicals were to be exchanged between cells, by ingestion or by cell division, the reactions within each cell must have tended to converge, to become compatible. The specific toxins and poisons by which some living forms today protect themselves must have been a relatively late, rare and specialized mechanism. Evolutionary stable strategies, to use John Maynard Smith's term, would tend to develop even in the absence of replicating molecules; they are a necessary homeodynamic property of the super-organismic network of living systems as a whole. Truly, we are all molecularly interdependent.

SOURCES OF ENERGY

Even before the problem of accurate replication had been resolved, there would have been another more pressing problem, that of energy. Such replication, as opposed to the mere splitting of membrane-bounded droplets, requires the synthesis of nucleic acids and proteins. The synthetic reactions that produce such macromolecules require an energy input (they are called *endergonic* reactions). The energy to drive them, excluding the special circumstances of electrical storms and volcanic eruptions, can come only by coupling the reactions to other, energy-releasing (*exergonic*) ones. I have earlier pointed to the contrast between today's molecular biologists, with their intense focus on the role of information in living systems, and those biochemists who flourished in the period prior to Watson and Crick and who were concerned with the problems of energy flow. Early life forms, proto-cells concentrated in liposomes, or even encrusted on the surface of catalytic clays, may have been able to absorb into themselves abiotically synthesized carbon- and nitrogen-based chemicals, but these abiotic stores would eventually have been exhausted, and an evolutionary bottleneck would have prevailed until the energy-generation problem could be solved. This must have happened either simultaneously with or before the emergence of reliable replicative mechanisms.

Today's living forms are divided into two broad categories: those which can obtain their energy by tapping into non-living sources (*autotrophes*), and those which require it prepackaged, supermarket style, in the form of convenience molecules like sugars or fats (*heterotrophes*). As discussed in Chapters 5 and 6, sugars and fats can be broken down – oxidized – to carbon dioxide and water. These are exergonic reactions, and the stepwise way in which they occur in the cell means that the energy is released in a sufficiently controlled manner that it can be trapped by using it to synthesize ATP, to be employed in its turn for a range of cellular activities from the synthesis of proteins and nucleic acids to muscle contraction and nervous transmission. Even for autotrophes, the most convenient prepackaged energy store comes in the form of sugar or fat molecules, so their first use of the abiotic energy they obtain is to synthesize sugars from carbon dioxide absorbed from the atmosphere. Heterotrophes can of course then live off this trapped energy, by eating either the autotrophes or other heterotrophes which have themselves eaten the autotrophes.

So one very early step in the history of life on Earth must have been the development of autotrophic energy-trapping mechanisms. A variety of such mechanisms are in principle available, on the basis of simple thermodynamic considerations and available chemistry; some indeed are still in use by specialist life forms living on the margins of volcanic, sulphur-rich lakes. But the most common, universally available source of energy is that derived from the Sun's radiation, and mechanisms of photosynthesis to exploit it must have been a crucial evolutionary step. Today's green plants have cells containing sophisticated systems for trapping solar energy, encapsulated in the intracellular organelles called chloroplasts. Hence the attraction of Lynn Margulis's suggestion that chloroplasts are the evolutionary descendants of once free-living photosynthesizing bacteria, which swapped their independence for the symbiotic security of multicellular life. Such a species-merging must have followed, not preceded, the appearance of DNA-based replication, as chloroplasts, like mitochondria, contain their own residual DNA. And, also as with mitochondria, far from being 'selfish' in Dawkins' sense, these primitive chloroplast genes must have been 'self-sacrificial', prepared to submerge their individual propagative rights in the interests of the organism, the proto-chloroplast, itself. Thus the cooperative symbiogenesis by which life as we know it today must have evolved provides an important alternative perspective to the ruthlessly individualistic competitive metaphor which underlies the ultra-Darwinist, replicator's-eye view of the world.

AT LAST, THE REPLICATORS ARRIVE!

So, some time after the development of effective mechanisms for generating and utilizing energy, though presumably before the development of the modern cellular systems of chloroplasts and mitochondria, replication based on nucleic acid would have emerged. The synthesis of simple nucleic acids has itself been achieved in the abiotic test-tube experiments I have already described, and once they had been incorporated within the metabolic web of the cell, they would offer a whole new range of properties. For they would now achieve a level of fidelity in copying and reproduction which would have been unobtainable by mere random division without them. For the reasons already advanced, it seems likely that RNAs, which are simpler molecules, would have appeared before DNAs, and because RNA molecules can show catalytic properties, the original enzymes might have been not protein- but RNA-based – ribozymes, in fact. This scenario, as I have said, has been christened 'RNA-world' by origin-of-life theorists.

Once nucleic acid – perhaps ribozyme-containing – cells had arisen, they would contain within themselves the rudiments of a faithful copying mechanism, an ability which so far as is known is today exclusively a property of the nucleic acids. Just how this mechanism settled down into its present-day form, based as it is on the trinity of DNA, RNA and protein, is a matter of intense speculation. At one point it was thought that there were particular conformational reasons – that is, resulting from their three-dimensional shapes – which would explain why any particular triplet of bases in the RNA molecule recognized a particular one of the twenty naturally occurring amino acids, but this idea has now been abandoned. Contingency, rather than laws of molecular form or adaptation, may rule at this point in the story. Once a particular set of nucleic acid–amino acid correspondences had emerged, convergence within the web would have been likely to help ensure its universality. In any event, the essential point is that, once cells operating these mechanisms had arisen, they would rapidly multiply and swamp all others, as only they could produce exact copies of themselves. Evolution, having generated nucleotide polymers within the primitive cells, had now also produced a mechanism which could be relied upon to amplify them, and before long to conquer the Earth – yet another reason why whatever the processes by which life forms were first generated, so far as life on this planet is concerned, they cannot repeat themselves. Just as organisms relying on the wrong optical isomers of amino acids or sugars are now excluded from emerging, so are those without a hi-fi replicative mechanism.

By this point in the story, with the development of faithful replicative systems and the energy-generating mechanisms with which to sustain them, life had arrived – by molecular biologists' definitions now as well as by mine. But in my version of the story, it had arrived without the help of implausible naked replicators, and with RNA, and later DNA, playing a proper, vital but not unilinearly determinist, role within the cell. Chickens, in this sense, came before eggs. It is in this sense that I have argued that life is inevitably autopoietic, self-generating, self-developing, self-evolving. The detailed routes that led from this speculative early replicative world to the present thirty million – or whatever – species are of course largely unknown and largely unknowable. We carry the history of that long trek inscribed in every cell within our bodies, and to make sense of ourselves we need to understand at least its outlines.

But in contrast to the molecular biologists' Bible, the Word, the nucleic acid script, was not in the beginning: it appeared later in the story, once there were already cells, organisms, prepared to receive and utilize it. Of course, once the Word had arrived, what followed can truly be said to be history if only because it can in some measure be read, like a book, from the periodically changing scripts inscribed in the mutating genomes of evolving organisms. But the script is merely a record. It does not in itself comprise the history of life, which is one of organisms, not of mere molecules.

NOTES

1. John 1.1–4.
2. Some researchers set the date even earlier, but the debate about such precision is outside my concerns here.
3. David S. McKay *et al.*, 'Search for past life on Mars'.
4. Alexander Oparin, *The Origin of Life on the Earth.* I was greatly excited when I first, as a young post-doc, read the Oparin hypothesis, and wrote about it warmly in my first book, *The Chemistry of Life*, in the 1960s. Later, the book was pirated into a Russian translation – this was before the then USSR signed the copyright convention – and the first I knew about it was the receipt of a copy of the Russian language edition inscribed by the then very elderly Oparin. My pride in this was only later shaken when Russian friends told me more of the negative part Oparin had played in the oppression and purging of Soviet scientists during the Lysenko period.
5. J. B. S. Haldane, 'The origin of life'.
6. Fred Hoyle and Chandra Wickramasinghe, *Lifecloud.*
7. Francis H. C. Crick, *Life Itself.*
8. Michael Behe, *Darwin's Black Box.*
9. R. J. P. Williams and J. J. R. Frausto da Silva, *The Natural Selection of the Chemical Elements.*
10. Stephen Mann and Geoffrey A. Ozin, 'Synthesis of inorganic materials with complex form'.
11. In *The Pasteurization of France*, Latour has criticized these famous experiments, regarded at the time as a final refutation of spontaneous generation, as set-piece displays rather than 'genuine', but the criticism is more of the method of argument, the rhetorical devices Pasteur used, than the conclusions.
12. James E. Lovelock, *Gaia: A New Look at Life on Earth.*
13. Lynn Margulis and Oona West, *Gaia and the Colonisation of Mars.*
14. Stanley L. Miller, 'A production of amino-acids under possible primitive Earth conditions'.
15. A. G. Cairns-Smith, *Seven Clues to the Origin of Life.*
16. Stuart Kauffman, *At Home in the Universe.*
17. Kauffman, 'Even peptides do it'.

10

The Poverty of Reductionism

[A man] with a rule and a pair of scales, and the multiplication table always in his pocket, sir, ready to weigh and measure any parcel of human nature, and tell you exactly what it comes to. It is a mere question of figures, a case of simple arithmetic . . . Time itself for the manufacturer becomes its own machinery: so much material wrought up, so much food consumed, so many powers worn out, so much money made.

Charles Dickens, describing Thomas Gradgrind, in *Hard Times*

THE RISE OF NEUROGENETIC DETERMINISM

It is time to shift gears. The preceding chapters of this book have taken issue with the fashionable gene's-eye view of biology which conceives living organisms as nothing but 'lumbering robots', assemblages of organs, tissues and chemicals created by and subject to the commands of a master-molecule whose goal is self-replication. By contrast, I have offered an alternative vision of biology which focuses instead on the autopoietic functions of organisms, their lifelines in space and time. In doing so I have been treating the issue of reductionism primarily as if it were an internal problem for biologists (and perhaps also for philosophers), about how to design and interpret experiments, and how to understand and explain living processes. I have tried to show why reductionist explanations are both so seductive and yet so inadequate in dealing with the complexities of the living world. I want now to move to the final and in some ways most polemical stage of the discussion, which I term *reductionism as ideology*. By this I mean the tendency, very marked in recent years, to insist on the primacy of reductionist over any other type of explanation, and to seek to account for very complex matters of animal – and above all of human – behaviour and social organization in terms of a reductionist precipice which begins with a social question and terminates with a molecule – often a gene. To return to my fable of the five biologists and the jumping frog, it is as if nothing else matters but the molecular biologist and the chemistry of actin and myosin.

The issues raised by these opposing visions are not confined to esoteric disputes between ivory-tower academics. I have emphasized the ideological power of modern biology in its claims to interpret and pronounce upon the human condition, to offer explanations and remedies for our social ills. From its Baconian inception, modern science has been about both knowledge and power, above all the power to control and dominate nature, including human nature. Nowhere perhaps has this Faustian pact been made so explicit as in the programme that has shaped molecular biology since its origins. Its very name was invented as long ago as the 1930s by Warren Weaver, of the Rockefeller Foundation, as part of a coherent policy by one of the major fund-givers in the field. That policy, drawing on prevalent eugenic thinking on the need to 'improve the race' by selective breeding, was specifically to achieve a 'science of man' which was also a science of social control.[1] As one of the early directors of the Foundation expressed it bluntly in 1934, its policies[2]

> are directed to the general problem of human behavior, with the aim of control through understanding. The Social sciences, for example, will concern themselves with the rationalization of social control; the Medical and Natural sciences propose a closely coordinated study of the sciences which underlie personal understanding and personal control . . . [and specifically] the problems of mental disease.

To this end, the Rockefeller concentrated its resources on the sciences of psychobiology and heredity, in the firm belief, fostered by Weaver, that such control would come about through the study of the 'ultimate littleness of things'. As I emphasized in the opening chapters of this book, how biologists – or any scientists – perceive the world is not the result of simply holding a true reflecting mirror up to nature: it is shaped by the history of our subject, by dominant social expectations and by the patterns of research funding. The sheer power and scale of the Rockefeller vision, backed as it was by its hundreds of millions of dollars, ensured that alternative understandings of biology withered. That was the fate, for example, of the 1930s Theoretical Biology Club in Cambridge, England, centred around Joseph Needham, whose non-reductionist approaches to metabolism, development and evolution were swept aside by the Rockefeller offer to fund an explicitly reductive biochemical research programme.[3]

Of course, the Rockefeller vision has been immensely productive in both scientific knowledge and technologies, the products of this Baconian alliance. Today we can see its lineage in the mushrooming biotechnology companies in the USA, Japan and Europe, in the 1990s Human Genome Project and Decade of the Brain. But to naturalize it as if it were the only way of understand the living world, and to ignore its explicit goals of social control and its implicit eugenic agenda, is to fail to grasp the directions in which it is leading us, as if modern science has simply transcended the ideologies that shaped it in the past. Today's molecular biology is, however unreflectingly, heir to this past, and cannot simply shrug it off. Thus the dramatic advances in knowledge of the past decades have been accompanied by ever more strident claims that the new genetics, molecular biology and neuroscience are about to explain, and in due course to modify, the human condition, and in doing so will usher in a new era of what some years ago one of the enthusiasts[4] for the new biology called a 'psychocivilized society':[5]

> . . . there should be tattooed on the forehead of every young person a symbol showing possession of the sickle-cell gene or whatever other similar gene . . . It is my opinion that legislation along this line, compulsory testing for defective genes before marriage, and some form of public or semi-public display of this possession, should be adopted.

The date of this quasi-Nazi proposal? Not the 1930s, but 1968. And its author? Hero of anti-war and alternative health movements, and twice Nobel prize-winner, once for chemistry and once for peace, Linus Pauling.

Week after week, newspapers report what are seen as major break-throughs in biological and medical understanding. Here's a random sampling: 'Stress, anxiety, depression: The new science of evolutionary psychology finds the roots of modern maladies in the genes' was the cover story for *Time* magazine for 28 August 1995. 'Gene hunters pursue elusive and complex traits of mind', claimed the *New York Times* on 31 October 1995. 'Studies link one gene to a specific personality', offered the *Talahassee Democrat* in January 1996. In July 1993 the London *Daily Mail* announced an 'Abortion hope after "gay genes" finding'. The London *Independent* carried an article entitled 'How genes shape the mind' (1 November 1995). More circumspectly, the London *Guardian* on 1 February 1996 described the hunt for 'intelligence genes' by Robert Plomin (newly appointed from the USA to a professorship at London's Maudsley Institute of Psychiatry) as 'the search for the clever stuff', and listed the 'losers in life's genetic lottery' as those who lack such genes.

Genes have been located, it is claimed, not only 'for' diseases like breast cancer but also 'for' homosexuality, alcoholism, criminality and a now notorious – and only half-facetious – speculation by Daniel Koshland, then editor of one of the world's premier scientific journals, *Science*, that there might even be genes for homelessness.[6] At the same time, drugs to extend life, improve memory or prevent 'compulsive shopping' make newspaper headlines. University scientists call press conferences, issuing promissory notes in which they claim to have discovered the biological causes of sexuality, or of violence in modern society. 'Twin studies suggest an even temperament may lie in the genes', claimed a press release from the University of Wisconsin in February 1994. A year later, the London-based medical charity the CIBA Foundation called a press conference to announce that they were sponsoring a closed meeting of behaviour geneticists whose research pointed to a 'biological' origin for the incidence of violent crime.[7]

The emerging synthesis of genetics and the brain sciences – *neurogenetics* – and its philosophical and political offspring, which we may call *neurogenetic determinism*, offers the prospect of identifying, ascribing causal power to, and eventually modifying genes which affect brain and behaviour. Neurogenetics claims to be able to answer the question of where, in a world full of individual pain and social disorder, we should look not merely to explain but, even more potently, to change our condition. While only the most extreme reductionist would suggest that we look for the origins of the Bosnian war in deficiencies in neurotransmitter mechanisms in Dr Radovan Karadzic's brain, and its cure by the mass prescription of Prozac, many of the arguments offered by neurogenetic determinism are not far removed from such extremes. Give the social its due, the claim runs, but in the last analysis the determinants are surely biological. And anyhow, we have some understanding and possibility of intervention into biological processes, by drugs, abortion or gene therapy, while by contrast – so such determinism insists – social interventions have been notoriously unsuccessful.

Urban violence, homelessness and psychic distress are desperately serious features of life in Europe and the USA today, and solutions must be sought. So the argument against hunting for neurogenetic explanations is not that it is immoral or unethical to do so. It is simply that, despite the seductive power of reductionism, neurogenetics is the wrong level of the disciplinary pyramid of Figure 1.1 (page 9) at which to find answers to many of the problems confronting us. It then becomes at best an inappropriate use of scarce human and financial resources, and at worst a substitute for social action. I need to reiterate this strongly if only because I find it so persistently, even perversely, misunderstood. I am distressed by the arrogance with which some biologists claim for their – our – discipline explanatory and interventionist powers which it certainly does not possess, and so cavalierly dismiss the counter-evidence.

TRUMPETING GENES

This is not a new debate. It has recurred in each generation at least since Darwin's day, most recently in the form of the polemical disputes over the explanatory powers of sociobiology in the 1970s and 1980s.[8] It is not my intention to go over that old ground again.[9] What is new today, however, is the way in which the mystique of the new genetics is seen as strengthening the reductionist argument. At its simplest, neurogenetic determinism argues for a directly causal relationship between gene and behaviour. A man is homosexual because he has a 'gay brain',[10] itself the product of 'gay genes',[11] and a woman is depressed because she has genes 'for' depression.[12] There is violence on the streets because people have 'violent' or 'criminal' genes[13]; people get drunk because they have genes 'for' alcoholism.[14] In a social and political environment conducive to such claims, and which has largely despaired of finding social solutions to social problems (although no one to my knowledge is researching the genetic 'causes' of homophobia, racism or financial fraud), these apparently scientific assertions become magnified by press and politicians, and researchers may argue that their more modest claims are traduced beyond their intentions – as with the disclaimers by Han Brunner concerning so-called 'aggression genes', to which I refer below. Yet this is hard to credit when the researchers themselves put so much effort into sales talk. The press releases surrounding the publication in 1992 of Simon LeVay's book *The Sexual Brain*,[15] which claimed, on the basis of his post-mortem studies of the brains of a number of gay men who died of Aids, to have located a specific region of the brain which differed in presumed gay from presumed straight men, or Dean Hamer's research paper in 1993 which claimed to have identified a 'gay gene',[16] were couched in language that left little need for media magnification.

The undoubted successes of molecular biology since the discovery of the double-helix structure of DNA in 1953 have fostered the sort of gung-ho triumphalism among geneticists not seen in science since the heydays of physics in the 1920s and 1930s – the belief that their science can explain everything that is to be explained about the human condition, and indeed can rebuild humanity in an improved image if allowed: 'give me a gene, and I can move the world'. Nor has biology hitherto been well served by philosophers, more accustomed to delivering critical analyses of the meta-claims of simpler sciences like physics. It is as if they have been bemused by the sheer encroachment of the claims of biology upon their very status as thinkers. Physics, after all, never proposed to colonize philosophy, but merely to live in harmony with it. The opening paragraphs of Wilson's *Sociobiology*, by contrast, make just this claim for the new biology, rendering redundant such human sciences as sociology, economics, politics and psychology. In response, many philosophers have retreated, while some have mutated into a new breed, of so-called bioethicists, pondering the moral dilemmas apparently opened up by the futures which biology – or at least genetics – seems to offer. Yet even this space is to be denied to philosophers, for the new molecular biologists want not merely to do their science but also to control its uses. Wilson, for instance, advocates a code of ethics which is 'genetically accurate and hence completely fair'.[17] Few professional philosophers seem prepared to subject these ethical claims to rigorous analysis – Mary Midgley being one honourable exception.[18] Today's buzzword is universal (ultra-)Darwinism.

REDUCTIONISM AS IDEOLOGY

Claims to explain phenomena as diverse as sexual orientation, mental distress, worldly success as measured by school performance, job or income, and violence on the streets of our major cities are scarcely minor concerns. We all want to know where to look to explain our personal successes and failures, our foibles and vices, to say nothing of the chronic crises we see in the world around us. For such problems we have the choice of invoking either social or personal explanations. If social, we can seek solutions through social action – improving the economy, changing the law, or working to alter the social structures of power and privilege. If personal, we can explore our own individual life history by way of psychotherapy. Or we can invoke the biological and claim that the roots of the problem we confront lie within individual brain structure, biochemistry or genetics. If the causes of our pleasures and our pains, our virtues and our vices, lie predominantly within the biological realm, then it is to neurogenetics that we should look for explanation, and to pharmacology and molecular engineering that we should turn for solutions.

As I have repeatedly emphasized, this simplification, with its implication that the world is divided into mutually incommensurable realms of causation in which explanations are either social *or* 'biological', its cheaply seductive dichotomies of nature *or* nurture, genes *or* environment, is fallacious. The phenomena of life are always and inexorably simultaneously about nature *and* nurture, and the phenomena of human existence and experience are always simultaneously biological *and* social. Adequate explanations must involve both.[19] Of course, for any serious natural scientist to deny the relevance of the social in favour of the biological would be equivalent to politicians denying that they were giving priority to party, rather than national interests; we are all interactionists now. In any search for explanation and intervention it is necessary to seek the appropriate level which effectively determines outcomes. Yet again and again one finds the reductionist claim, unqualified, making the headlines and setting the research agenda.

Neurogenetic determinism, I argue, is based on a faulty reductive sequence whose steps include reification, arbitrary agglomeration, improper quantification, belief in statistical 'normality', spurious localization, misplaced causality, dichotomous partitioning between genetic and environmental causes, and the confounding of metaphor with homology. As will become clear, no individual step in this sequence is inevitably in error, it is just that each is slippery and the danger of tumbling very great. The issue at stake here is not so much the formal philosophical one which I addressed in Chapter 4, but the question of the appropriate level of organization of matter at which to seek causally effective determinants of the behaviour of individuals and societies. The structure of the argument is similar whether the discussion focuses on intelligence, sexuality or violence, and I shall base my analysis mainly around these themes.

REIFICATION

The first step in the process is *reification*. Reification converts a dynamic process into a static phenomenon. Violence is the term used to describe certain sequences of interactions between persons, or even between a person and their non-human environment. That is, it is a process. Reification transforms the process into a fixed thing – *aggression* – which can be abstracted from the dynamically interactive system in which it appears and studied in isolation, as it were, in the test-tube. This is the thinking that has led to regarding aggression as a phenotypic character, to be analysed by the modern counterparts of Mendelian methods. In Chapter 5, I pointed to the difficulties inherent in regarding even apparently straightforward aspects of an individual, such as the colour of a pea or an eye as a unitary 'character'. To regard an aspect of behaviour as an isolable character is much more problematic. In Chapter 2, I described the care required, even within the methodologically reduced framework of an ethogram, in abstracting and defining the behaviour of a single individual held in relative isolation. Yet if the activity described by the term 'violence', or 'altruism', or 'sexuality', can be expressed only in an interaction between individuals, to reify the process and pretend that it is in any sense a character that can be isolated is to lose its meaning. It is to consider the frog jump without taking the snake into account.

ARBITRARY AGGLOMERATION

Arbitrary agglomeration carries reification a step further, lumping together many different reified interactions as if they were all exemplars of the one character. Thus *aggression* becomes the term used to describe processes as disparate as a man abusing his lover or child, fights between football fans, strikers resisting police, racist attacks on ethnic minorities, and civil and national wars. Agglomeration proceeds by assuming each of these social processes to be merely a reified manifestation of some unitary underlying property of the individuals, so that identical biological mechanisms are involved in, or even cause, each. This is well illustrated by a research paper, published in *Science* in 1993, by a team led by Han Brunner.[20] It described a Dutch family (*pedigree* is the technical term), some of whose menfolk were reported as being abnormally violent; in particular, eight men 'living in different parts of the country at different times' across three generations showed an 'abnormal behavioural phenotype'. The types of behaviour included 'aggressive outbursts, arson, attempted rape and exhibitionism'. Can such widely differing types of behaviour, described so baldly as to isolate them from their social context, legitimately be subsumed under the single heading of aggression? It is unlikely that such an assertion, if made in the context of a study of non-human animal behaviour, would pass muster (I certainly couldn't get away with reporting a study involving such varied behaviour in eight chicks!). Yet Brunner's paper was published in one of the world's most prestigious journals, with considerable attendant publicity. (Parenthetically, it is interesting how many of these rather sensationalist and often scientifically dubious papers claiming the identification of specific gene-based causes for human problems have been published in *Science*. The journal's rival, *Nature*, has been much more circumspect.)

The paper attracted much attention by reporting that each of these 'violent' individuals also carries a mutation in the gene coding for the enzyme monoamine oxidase (MAOA) which, among other functions, is associated with the metabolism of a particular neurotransmitter and is believed to be site of action of a number of psychotropic drugs. Could this mutation then be the 'cause' of the reported violence? Brunner himself subsequently disclaimed the direct link, and indeed, dissociated himself from the public claims that his group had identified a 'gene for aggression', claiming that this was merely a journalistic distortion.[21] Yet the claim is now widely cited in the research literature, in which what Brunner's paper described in its title as 'abnormal' now becomes 'aggressive' behaviour. Thus a paper whose title commenced with these two words, describing mice lacking the monoamine oxidase A enzyme, appeared in *Science* two years after the Brunner paper. The authors, a primarily French group headed by Olivier Cases, described the mouse pups as showing 'trembling, difficulty in righting, and fearfulness . . . frantic running and falling over . . . [disturbed] sleep . . . propensity to bite the experimenter . . . hunched posture . . .'.[22] Of all these features of disturbed development, the authors chose to include only 'aggression' in their paper's title, and to conclude their account by claiming that these results support 'the idea that the particularly aggressive behavior of the few known human males lacking MAOA . . . is a more direct consequence of MAO deficiency'. When I pointed out, in a letter to *Science*, that what the Cases paper headlined as aggression was a minor and scarcely surprising aspect of this grossly disturbed developmental pattern, one of the authors telephoned me to explain that they had highlighted aggression this way because it seemed the best way of drawing attention to their results.

More disturbingly, this type of evidence, slight though it may seem, has become part of the arsenal of argument employed, for example, by the US Federal Violence Initiative, aimed at identifying inner city children regarded as 'at risk' of becoming violent in later life as a result of predisposing biochemical or genetic factors. This programme, proposed originally by the then director of the US National Institute of Mental Health, Frederick Goodwin, originally ran into a hostile barrage of publicity over its potentially racist overtones, with its repeated coded references to 'high-impact inner city' youth. Not long afterwards Goodwin left his directorship, and plans to hold a meeting to discuss his proposals were several times abandoned.[23] None the less, parts of the research programme have been implemented in the USA, particularly in Chicago.[24]

As with each step in the reductionist cascade I am describing, the problem does not lie in the fact that, as researchers, within the methodology available to us, we need to classify – to group together different types of observation as having something in common. These are not inevitably illegitimate steps, as I argued previously in terms of my own studies of chick pecking as exemplifying memory. Science seems often to proceed by alternately grouping together different phenomena as aspects of the same (lumping) and recognizing differences between them (splitting). However, lumping arson and exhibitionism together in the same category as both examples of the 'natural kind' called 'violence' is not likely to make much sense to either a criminologist or a judge and jury in court.

To get round this difficulty, some researchers have recently relabelled such cases so that they no longer appear as examples of 'violence', but of a different category, of 'antisocial behaviour' now also regarded as a natural kind.[25] Far from solving the problem, such relabelling only makes it worse. Just as agglomeration lumps together disparate activities, so the identical act may be regarded as socially acceptable or unacceptable depending on the circumstances in which it is carried out. Bombing a government building in enemy territory if you are a pilot and your nation is at war is socially praiseworthy; on the other hand, if you are a member of the society whose buildings you bomb you are guilty of the antisocial behaviour called terrorism. Contrast the medals given to US pilots during the Gulf War with the criminal charges against the bombers of the Federal office building in Oklahoma City. Perhaps the clearest-cut example comes from an episode in Northern Ireland in 1990. A British soldier, Lee Clegg, was on duty at an army checkpoint when a stolen car crashed through the roadblock. Private Clegg lifted his rifle and shot dead one of the occupants of the car, a teenage girl who had been joyriding. He was charged, and convicted of murder, perhaps the ultimate in antisocial behaviour. The army, supported vociferously by the English tabloid press, was outraged and waged a vigorous and ultimately successful campaign for his release and reinstatement. He was, they argued, doing his duty, the car might after all have held IRA terrorists, not joyriding teenagers – in which case he might even have been given a medal. By 1997 he had been promoted to lance-corporal, and was seeking compensation for wrongful arrest and imprisonment. So the identical act can be defined either as socially approved or antisocial, depending now not on the act itself but on the perception of those who observe it. How can this conceivably form the basis for a biological, individually based categorization, in which we look for unusual genes for neurotransmitter enzymes in Lee Clegg's brain to explain what has happened? Antisocial behaviour is clearly not a natural kind.

IMPROPER QUANTIFICATION

Improper quantification argues that reified and agglomerated characters can be given numerical values. If a person is violent, or intelligent, one can ask how violent, how intelligent, in comparison with other people. This assumption, that any phenomenon can be measured and scored, reflects the belief, to which I have already referred, that to mathematicize something is in some way to capture and control it. The best-known example is the use of the IQ (intelligence quotient) scale to describe and measure intelligence. Along with many others, I have written previously about the history of this scale and some of the fallacies embedded within its use, and there is no need to repeat these arguments in detail here.[26]

The first steps involve reifications and agglomerations which parallel those described above for violence. 'Intelligent behaviour', essentially an interactive process between an individual and others, or with the social, living and inanimate worlds, becomes fixed as a unitary character. Many different examples of such behaviour are then all taken to be manifestations of something called, as if finally to freeze dynamics into statics, 'crystallized intelligence', and given a special symbol, g, originally introduced by the psychologist Charles Spearman in the 1920s (is it only coincidence that this is also the symbol for one of the most hallowed of physical forces, that of gravity?). Tests are then devised to measure this inferred hidden constant. Subjects are asked a series of questions, supposedly not dependent on school education, class or culture, but instead assessing underlying absolute skills, such as matching patterns or identifying logical sequences of numbers or words. The subject's score on these tests is then compared with that for the general population (or, for children, others of the same age group), and the resulting comparative figure is called the IQ. Of all the assumptions built into this process, for the moment I want to consider only one: the extraordinary belief that the multiple aspects of behaviour (even reified and agglomerated behaviour) that contribute to what we may recognize as intelligence – speed and accuracy of responding to new information, skill at deriving meanings from ambiguous social situations, capacity to innovate in novel environments, and many others as well – can all be reduced to a single number, so that the entire human population can be ranked by it, just as they might be if we were to line them all up by height.

Of course, to achieve this type of mathematical reduction it is necessary to discount many of these richly interacting human capacities, despite the fact that to most people they would seem to be among the most salient aspects of what is called intelligence. Instead, such psychometricians retreat into a private world inhabited only by like-minded devotees of the art of counting. Indeed, they find it difficult to relate to other brain and behavioural scientists, who mostly look askance at psychometry's commitment to arbitrary numerology. (In practice this means that the only other discipline to which they can relate, and with which psychometry has historically been linked, is a certain subarea of behaviour genetics. Indeed the two, psychometry and behaviour genetics, are the twin offspring of the eugenic movements of the early twentieth century.[27]) To see this cavalier rejection of anything other than the reduction of intelligence at its most arbitrary, one need go no further than the first chapter of Herrnstein and Murray's *The Bell Curve*, which, faced with the voluminous critiques, from many different perspectives, of such reduction of intelligence to a single score, sweeps aside all opposition. Intelligence, they insist, is not to be confounded with talent, insight, creativity, or capacity to find or solve problems or resolve difficulties, any more than it has anything to do with musical, spatial, mathematical or kinaesthetic ability, sensitivity, charm or persuasiveness:[28]

> There is such a thing as a general factor of cognitive ability on which human beings differ.
>
> All standardized tests of academic aptitude or achievement measure this ability to some degree, but IQ tests expressly designed for that purpose measure it most accurately.
>
> IQ scores match, to a first degree, whatever it is that people mean when they use the word *intelligent* or *smart* in ordinary language.

Thus intelligence is what intelligence tests measure and if other tests, constructed on different principles, fail to conform by providing a measure compatible with this unitary view of g, they are simply dismissed as being beneath consideration.

STATISTICS AND THE NORM

Belief in *statistical normality* assumes that in any given population the distribution of such behavioural scores takes the form known as a Gaussian distribution, the famous bell-shaped curve (Figure 10.1).

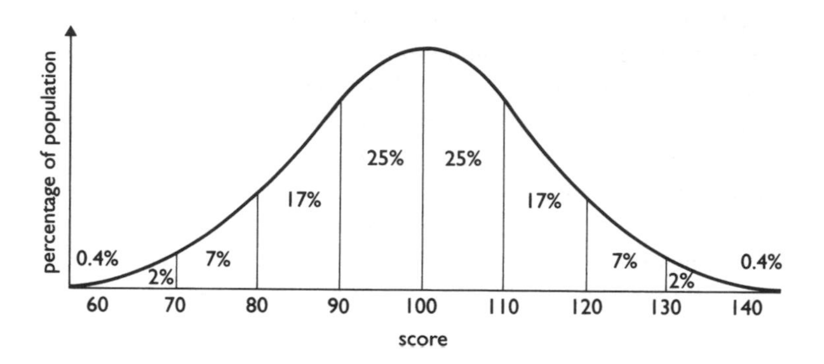

Figure 10.1 The bell curve.

This is known to statisticians as a 'normal' distribution. One of the best-known examples of its application is with IQ, the tests for which successive generations of psychometricians refined and remoulded until their results (almost) fitted the approved statistical shape. That is, tests which did not result in distributing the population according to the curve were rejected, or test items within them modified, until they fitted the curve, a feat achieved between the wars in the various revisions of what became known as the Stanford–Binet IQ test, originally developed in the 1920s. The curve-fitters also ran into another problem. When they looked at how males and females (boys and girls) performed on the tests, girls outperformed the boys on certain items, thus recording an apparently higher IQ. As the testers assumed that there should be no sex differences in IQ scores, items on which the two sexes scored differently were again adjusted until, on average, there were no longer any differences between them. However, when the tests showed average differences in score between people from working and middle classes, or between blacks and whites, these were assumed to reflect 'real' underlying differences in intelligence. It is in fact possible to construct tests on which, for instance, working-class children score higher than middle-class, but these are discounted. My late colleague Brian Lewis did this by arguing that working-class children had to cope with much more 'disinformation' – lies – than did middle-class children. He designed a test in which schoolchildren had to sort out strategies from a mixture of true and misleading statements. Working-class children did much better on these tests. (Modern testers sometimes employ so-called culture-fair tests, ignoring the fact that these have been standardized pa10calready against the Stanford–Binet and so are likely to perpetuate any biases implicit in the earlier tests.)

This procedure demonstrates how the ideological commitments of the testers can serve to construct a biology which they then assume they have simply read off from nature. But worse is the assumption that the entire population can be distributed along a single dimension, which is to confuse a biological phenomenon with a statistical manipulation. There is no biological necessity for such a one-dimensional distribution, nor for one in which the population shows such a convenient spread. It is perfectly possible to set examinations in which virtually everyone scores 100 per cent; the British university penchant for 10 per cent first class degrees, 10 per cent thirds and 10 per cent fails, with everyone else comfortably in the middle, with a second, is a convention, not a law of nature (Figure 10.2).

Yet the power of this reified statistic should not be underestimated. It conveniently conflates two different concepts of 'normality'. The statistical sense of the term does not have a 'value' attached to it: 'normal' merely describes a particular shape of curve which has the property that 95 per cent of its area is to be found within a defined distance – two standard deviations – of the mean. But in common parlance the term does indeed mean 'normative'. It describes not merely how things are, but how they ought to be: to lie more than

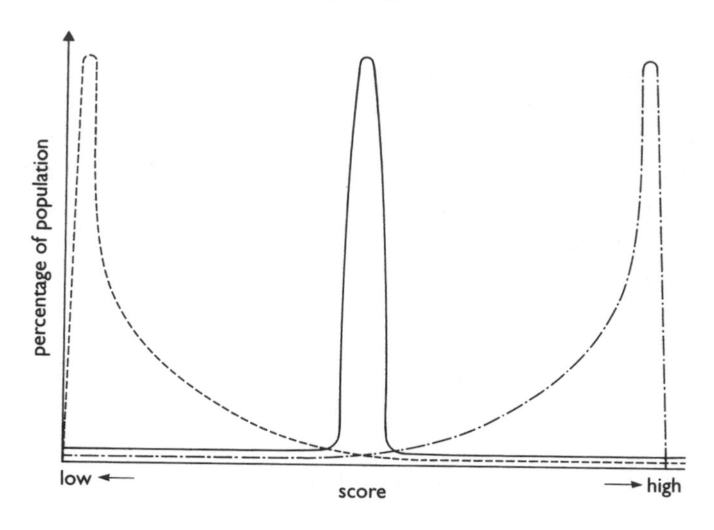

Figure 10.2 Possible distribution of test scores within a population. Three non-bell-curve potential distributions are shown; any one of them, and any others in between, are possible, depending on the design of the test.

two standard deviations from the mean in a Gaussian distribution is to be abnormal, with all that this implies. When Herrnstein and Murray called their book *The Bell Curve*, they played precisely into these multiple meanings of reified normality.

SPURIOUS LOCALIZATION

Having reified processes into objects and arbitrarily quantified them, the reified object ceases to be a property even of the individual, but instead becomes a property of part of the individual. Hence the penchant for speaking of, for example, schizophrenic brains, genes – or even urine – rather than of brains, genes or urine derived from a person diagnosed as suffering from schizophrenia. Of course, everyone ought to know (and does, at least on Sundays, just as everyone is an interactionist these days) that this is a shorthand, but the resonance of 'gay brains' or 'selfish genes' does more than merely sell books for their scientific authors: it both reflects and endorses the modes of thought and explanation that constitute neurogenetic determinism, for it disarticulates the complex properties of individuals into isolated and localized lumps of biology.

Thus recent years have seen an unusually polemical debate, more reminiscent of the early days of nineteenth-century phrenology than of modern research, among different neuroanatomists each claiming to have found 'the' brain seat of homosexuality. Two regions in particular have been in contention for the honour of conveying male homosexual preference: one the great band of nerve fibres that connects the two halves of the brain, the corpus callosum, the other a cluster of nerve cells deep in the brain called the hypothalamus. According to Laura Allen[29] in California, men and women have corpora callosa, which if measured at a particular angle, differ in thickness, while gay men are, naturally, intermediate between proper heterosexual males and females. By contrast, Dick Swaab in Amsterdam[30] and Simon LeVay in La Jolla, California,[31] focus on the hypothalamus, each proposing a different part of this complex structure as differing in volume between gay and presumed straight men. LeVay's study made the headlines partly because he used autopsy material from men who had died of Aids, partly because he followed his research paper up with a popular book, *The Sexual Brain*, and partly because he is himself a declared gay. Indeed, he argued that the finding of a site in the brain for gayness was liberatory because it relieved men of the stigma of immorality, and would alleviate the fear expressed by some in the straight community that they could catch this sexual 'disease' by the contagion of mixing in the wrong company.

I don't want to go into a detailed analysis here of the empirical evidence offered by the three neuroanatomists, or by the geneticist Dean Hamer, of the National Institutes of Health in Bethesda, Maryland, who in 1993 trumped the anatomical studies by reporting that he had found not a gay brain, but a marker for a 'gay gene'.[32] These studies have recently been subjected to detailed and stringent empirical criticism by Anne Fausto-Sterling,[33] and there have apparently been problems in replicating Hamer's findings in other samples. My concern here is once more with the structure of the argument deployed by those seeking to locate homosexuality in a bit of the brain or an aberrant gene, for it shows all the features I have already described for attempts to home in on violence and intelligence, and more besides. The expression of same-sex preference is scarcely a stable category, either within an individual's lifetime or historically – indeed, that 'homosexual' might be used as a term to describe an individual, rather than part of a continuum of sexual activities and preferences available to all, seems to have been a relatively modern development.[34] What the reductionist argument does is to remove the description of sexual activity or preference from being part of a relationship between two individuals, reify it and turn it into the phenotypic 'character' resulting from one or more abnormal, gay genes. As always, it deprives the term of personal, social or historical meaning, as if to engage in same-sex erotic activity or even to express a same-sex preferred orientation meant the same in Plato's Greece, Victorian England and San Francisco in the 1960s.

Just as homosexuality is 'located' in the hypothalamus, so aggression had been 'located' in another set of structures within the brain, the limbic system, and in particular one part thereof, the amygdala. In the 1970s two psychosurgeons proposed to treat inner-city violence by 'amygdalectomizing' militant ringleaders from America's inner-city ghettos[35] – that is, by cutting out the offending region, rather as in the biblical exhortation to pluck out your eye if it offends you. I used to believe that things were a little more sophisticated today, but a 1995 television documentary persuaded me otherwise. It showed California-based psychologist Adrian Raine standing in front of two brain images taken by PET (positron emission tomography) scanning, and explaining that one, the brain of 'a murderer', showed 'low activity' in the frontal cortex by comparison with the other, the 'normal' brain.[36] I gloomily concluded that the days of the nineteenth-century Italian criminologist Cesare Lombroso, who believed you could tell thieves, murderers and swindlers apart by the shapes of their heads, were not that long past.

Raine was theorizing that the function of the 'more evolved' cortex in humans was to control the 'primeval' limbic system, and that where frontal activity is low the amygdala and other limbic systems are out of control, and left to their own devices will drive their owners to violence. It is not made clear whether a similar finding would apply to scans of the brains of the war heroes who have been responsible for some of the greatest massacres of modern times, Stormin' Norman and the killings of fleeing Iraqi troops on the Basra Road in 1991, or Ratko Mladic and the mass graves of the Muslim men of Srebenica in 1995. What is certain is that a view of the brain as composed of 'less' and 'more' evolved structures is yet another of those evolutionary fantasies. It is of course species, not individual parts of an organism, which evolve, and during such evolution old structures acquire new functions. The great mass of the cerebral cortex, in humans and other mammals, shows evolutionary descent from the olfactory bulb, still there in present-day reptiles. But that doesn't mean that we think by smelling.

Raine's claims take us back to an older tradition, of 'localizing' reified properties. More frequently these days, that localization takes the form not of a brain structure, but of an abnormality in some brain chemical – a neurotransmitter or an enzyme or the gene responsible for its production. The particular substance in question tends to fluctuate with the fashionable molecule of the moment. Thus a few years ago much attention was paid to one particular neurotransmitter, the substance gamma amino butyric acid (GABA) as being particularly associated with aggressive behaviour. Today, aggression is more likely to be explained as being 'caused' by a disorder of the metabolism of serotonin (specifically, the re-uptake of the secreted neurotransmitter into the cells of the brain). Abnormalities of serotonin re-uptake mechanisms are blamed for everything from depression and suicide to 'impulsive behaviour' and violence; the universal panacea is Prozac, one of a family of drugs which selectively inhibits serotonin re-uptake.[37]

MISPLACED CAUSATION

It is at this point that neurogenetic determinism introduces its misplaced sense of causality. It is of course probable – indeed in some contexts certain – that during aggressive encounters people show dramatic changes in, for instance, the levels of steroid hormones and adrenalin circulating in their bloodstream and the release of neurotransmitters in their brain, all of which can be affected by drug treatments. People whose life history includes many such encounters are likely to show lasting differences in a variety of brain and body markers. But to describe such changes as if they were the *causes* of particular behaviours is to mistake correlation, or even consequence, for cause. When you have a cold, your nose runs. Yet despite the invariable correlation of the two, it would be a mistake to believe that the cold was caused by the nasal mucus; the chain of cause-and-effect runs in the reverse direction. Nor, despite the fact that Prozac both inhibits serotonin re-uptake mechanisms and may diminish the likelihood of you committing suicide or murder, does this mean that the level of serotonin release in your brain is the cause of your desire to kill yourself – or someone else. After all, when one has toothache one can alleviate the pain by taking aspirin, but it does not follow that the cause of the toothache is too little aspirin in the brain.

This misconception (which follows the logic of the biochemist who argues that the cause of the frog jumping lies in the chemistry of actin and myosin) has for decades dogged the interpretation of the biochemical and brain correlates of psychiatric disorders,[38] yet it still continues. Thus recent claims that an abnormality in the receptor molecules for yet another neurotransmitter, dopamine, could underlie susceptibility to substance abuse were countered by the argument that the abnormality was the result, not the cause, of drug taking.[39] Such beliefs are an almost inevitable consequence of the processes of reification and agglomeration, for if there is one single thing called, for instance, alcoholism, then it is considered appropriate to seek a single causative agent.

DICHOTOMOUS PARTITIONING

If aggression, or antisocial behaviour, or homosexuality are 'caused' by some 'abnormality' in brain structure, or in biochemistry, or by some hormonal imbalance, what in turn 'causes' these? They could of course be the consequences of some feature in the environment (and those who believe so usually argue that they result from some aspect of early rearing or poor diet, as when infant 'temperament' in early months is claimed to predict later poor performance at school or adult violence[40]). More often, though, attention turns to those well-known first causes, the genes, and the apparatus of heritability studies is wheeled out. For even if there is difficulty in regarding such socially defined attributes as characters in the Mendelian sense, if they correlate with a 'real' measure such as the level of an enzyme or neurotransmitter, then the heritability of this can surely be determined. A good example of this mode of thinking is the claim that IQ scores correlate with a more neurophysiological measure referred to as 'inspection time', whose heritability can then be assessed. In Chapter 7 I took a detour through the history and mathematics of the heritability measure, and explained why – except in the very specific context for which it was originally devised (agricultural breeding experiments) – it was rarely applicable, widely misunderstood and in most cases meaningless. Sadly, this has not prevented behaviour geneticists and psychometricians from endeavouring to apply it; nor has it been deprived of its ideological resonance, as when claims are made that the heritability of intelligence – or rather of IQ test score – is as high as 80 per cent.

As I mentioned in Chapter 7, political orientation, neuroticism and attitudes to military drill, royalty, censorship and divorce, among many others, are all supposed to show relatively high heritability. Indeed, it becomes hard to find any human attribute or belief, even the most seemingly trivial, to which the heritability statistics fail to identify an apparently significant genetic contribution. New and sophisticated statistical techniques, such as the so-called quantitative trait locus analysis,[41] are employed which purport to show that even those conditions for which major genetic causation cannot be shown (Alzheimer's disease is a good example, where only about 5 per cent of the cases are clearly associated with a specific genetic dysfunction) are in fact the result of the small additive effects of many genes. And while no one claims that heritability equals destiny, nor that such an estimate provides information about any specific individual, rather than merely measuring the variance within a population, the whole tenor of the approach is none the less to transfer the burden of explanation, and if appropriate of intervention, from the social or even the personal level to that of pharmacological or genetic control.

CONFOUNDING METAPHOR WITH HOMOLOGY

If first causes are genetic, then the adaptationist paradigm within ultra-Darwinism must seek to account for how they may have evolved. It then becomes appropriate to seek equivalents of the human behaviour under consideration in the non-human animal world – that is, to find an animal model in which the behaviour can be more readily controlled, manipulated and quantified. Place an unfamiliar mouse in a cage occupied by a rat, and the rat is likely eventually to kill the mouse. The time taken for the rat to perform this act is taken as a surrogate for the rat's aggression; some rats will kill quickly, others slowly or even not at all. The rat which kills in thirty seconds is on this scale twice as aggressive as the rat which takes a minute. Such a measure, dignified as *muricidal behaviour*, serves as a quantitative index for the study of aggression, ignoring the many other aspects of the rat–mouse interaction, for instance the dimensions, shape and degree of familiarity of the cage environment to the participants in the muricidal interaction, whether there are opportunities for retreat or escape, and the prior history of interactions between the pair. And it is not that these are merely speculative variables, for many of them have been studied in detail by ethologists and shown to profoundly affect the nature of the relationships between the animals.

But the reductive procedure is taken further, for it is then assumed that, just as time to kill becomes a surrogate for a measure of aggression, so this behaviour in the rat is transmogrified into an analogue of the aggression shown by drive-by gangs shooting up a district in Los Angeles, as in the concluding sentences to the paper by Cases referred to above. That is, if one can find physiological or biochemical mechanisms – brain regions, neurotransmitters or genes – associated with the so-called 'aggression' in mouse-killing rats, then there should be equivalent or identical brain regions, neurotransmitters or genes involved in human aggression too.[42] Similar arguments are applied to the search for animal models for drug dependency and alcoholism.[43] This type of evolutionary fantasy at best confounds a metaphor or analogue with a homologue in the sense defined in Chapter 2, and this is why I have to be so careful in my own claims that memory in chicks is a homologue of memory in humans. At worst, it simply makes a false equation between different meanings of the word 'aggression'. But it has become the vital, ultimate link in the chain-mail armour of reductive ideology.

THE CONSEQUENCES OF REDUCTIONIST FALLACIES

From the birth of modern science, methodological reductionism has proved a powerful and effective lever with which to move the world. We owe to it many of the most penetrating insights into mechanisms in every field of science, including biology. But, especially in biology, complexity and dynamics, open rather than closed systems, are norms rather than exceptions, and the methodology of reductionism, however powerful, has difficulties in dealing with complexity – indeed, it may prove positively misleading.

Furthermore, as I argued in Chapter 4, reductionist methodology very easily tips over into reductionist philosophy. This philosophy, the 'nothing-buttery' that collapses all of science into physics, is untenable. Nor is it possible to retain a partial reductionism by which one can choose voluntarily to halt the descent from social behaviour to quantum physics at any convenient point. By its very nature, reductionism is all or none, while an eliminative reductive philosophy fails to account for the new meanings of phenomena which emerge at each successive level of organization of matter. The particular chemical properties of haemoglobin are essential to its significance as an oxygen-carrying molecule within the physiology of the organism, but this functional role cannot be reduced to simple chemistry, any more than the properties of the actin and myosin that enable a frog's muscle to contract can in themselves explain why the frog jumps when it sees a snake. Each level of organization of the universe has its own meanings, which disappear at lower levels. In short, we require epistemological diversity in order to understand the ontological unity of our world.

And so to reductionism as an ideology which insists on trying to account for higher-level phenomena in terms of lower-level properties. It does so by means of a faulty cascade of reification, arbitrary agglomeration, improper quantification, belief in normative statistics, spurious localization, misplaced causation and the confounding of metaphors with homologues. The motivations for such reductive explanations derive in part from the power of reductionism as both methodology and philosophy, but even more strongly from the urgent pressure to find explanations for the scale of social and personal distress in advanced industrial societies at the end of the twentieth century, explanations which shift the 'blame' for the problem away from the political realm and onto the individual. This drift from the social was memorably summed up by Margaret Thatcher during her years as Britain's prime minister, when she is said to have claimed that there is no such thing as society, only individuals and their families – thus fascinatingly rephrasing Watson's claim of 'only atoms'.

Reductionist ideology has a number of serious consequences. It hinders us biologists from thinking adequately about the phenomena we wish to understand. But two consequences at least lie in the social and political domain rather than the scientific, and need spelling out briefly here. First, reductionist ideology serves to relocate social problems to the individual, thus 'blaming the victim' rather than exploring the societal roots and determinants of the phenomena that concern us. Violence in modern society is no longer to be explained in terms of inner-city squalor, unemployment, extremes of wealth and poverty, and the loss of the hope that by collective effort we might create a better society. Rather, it is a problem resulting from the presence of individual violent persons, themselves violent as a result of disorders in their biochemical or genetic constitution.

But in a strange way, the blame is simultaneously placed upon them and lifted from them. Where once a murderer might have been regarded as morally culpable, or the cause of his (as it almost invariably is) violence sought in an unhappy or abused childhood, now it is argued to be due to lower 'frontal activity' or chemical imbalances in his brain, themselves the consequence of faulty genes or birthing difficulties. Thus, in a recent US court case the lawyer acting for a murderer, Stephen Mobley, sentenced to death for the violent slaughter of the manager of a pizza parlour, sought permission to mount a genetic defence against the sentence, claiming that his client may be endowed with the same mutation in his monoamine oxidase gene that Brunner found in the Dutch pedigree he studied. Mobley would not then be 'responsible' for the murder he committed: 'It was not me, it was my genes.'[44] Similarly, if homosexuality is 'in the genes', a gay man should not, in a homophobic society, be regarded as morally culpable, still less guilty of criminal behaviour, for following his genetic dictates. It is not surprising, therefore, that certain sections of the gay and lesbian community have actively welcomed the determinist claims of LeVay and Hamer, or that both the Christian fundamentalist right and the judiciary are worried about just how far the determinist argument can be stretched.

The second immediate social consequence of reductionist ideology is that attention and funding is diverted from the social to the molecular. If the streets of Moscow are full of vodka-soaked drunks, and rates of alcoholism are catastrophically high among native Americans or Australian aborigines, the ideology demands the funding of research into the genetics and biochemistry of alcoholism. And it becomes more productive to study the roots of violent 'temperament' in babies and young children than to legislate to remove handguns from society. The point is that, as the whole of my argument up till now has stressed, for any phenomenon in the living world in general, and the human social world in particular, one can offer multiple forms of explanation, of which the reductionist one, properly formulated, is legitimate. But for any such phenomenon there are also *determining levels* of explanation – those that account most clearly for the specificity of the phenomenon, and also indicate potential access-points for intervention into it.

Let me come back to violence again, and for the last time. Crimes of violence are more frequently carried out by men than by women (although this picture is changing in both the USA and Britain). One may argue that this says something about the Y chromosome, carried by men and not women, but the overwhelming majority of men are not violent criminals, so the policy implications of research seeking to study the Y chromosome in the context of crime – short of selective abortion of all male foetuses – are negligible. Violent crime is much higher in the USA than in Europe – higher, for instance than in Britain, and much higher than in Sweden. Could this be accounted for by some unique feature of the American genotype? Well, possibly, but pretty unlikely, since much of the American population originated by migration from Europe. But also the rates of violent crime change dramatically over quite short time periods. For instance, the death rate from homicide among young US males increased by 54 per cent between 1985 and 1994. No genetically based explanation can account for this increase, so it becomes more helpful to ask instead what has changed in the USA over this period which might account for such an increase. What is different about the organization of US society from that of Europe? Could one important difference be the estimated 280 million handguns in personal possession in the USA? Unlike reductionist ones, such hypotheses may provide pointers for meaningful intervention.

So while *of course* it is axiomatic that there is something different about the biochemical and physiological state of someone who is in the process of committing a murder from those states in the same person when he is in a prison cell, and probably between the murdering individual and someone who in similar circumstances does not murder, this difference cannot be relevant to answering questions about the causes and responses to social violence. Nor, therefore, can it represent the appropriate level at which to intervene if we wish to reduce the amount of violence on the streets. A programme devoted to determining what levels of serotonin might predispose a person to an increased statistical possibility of engaging in one of a number of activities, from suicide through depression to murder, followed by the mass screening of individual children to identify individuals at risk, their drugging throughout life, and/or their raising in environments designed to alter their serotonin levels – which is, after all, the action programme that would result from an attempt to define the genetic/biochemical as the right level for intervention – has only to be enunciated to demonstrate its fatuity. Good, effective science requires a better recognition of determining explanation, and hence of the determining level at which to intervene. Failing this, it becomes a waste of human ingenuity and resources, a powerful ideological strategy of victim-blaming and a distraction from the real tasks facing both science and society.

NOTES

1. Lily E. Kay, *The Molecular Vision of Life.*
2. Mason, speaking in 1934, quoted by Kay, *The Molecular Vision of Life*, p. 46.
3. Donna Haraway, *Crystals, Fabrics and Fields.*
4. J. M. R. Delgado, *Physical Control of the Mind.*
5. Linus Pauling, 'Reflections on the new biology', p. 269, quoted in Kay, *The Molecular Vision of Life*, p. 276.
6. Daniel Koshland, 'Editorial'.
7. Later published as: Gregory Bock and Jamie Goode (eds), *Genetics of Criminal and Antisocial Behaviour.*
8. Edward O. Wilson, *Sociobiology: The New Synthesis*; Philip Kitcher, *Vaulting Ambition.*
9. Steven Rose, Richard C. Lewontin and Leon Kamin, *Not in Our Genes.*
10. Simon LeVay, 'A difference in hypothalamic structure . . .'.
11. Dean Hamer and P. Copeland, *The Science of Desire.*
12. David B. Cohen, *Out of the Blue.*
13. A. Reiss and J. Roth, *Understanding and Preventing Violence.*
14. M. Galanter (ed.), *Recent Developments in Alcoholism.*
15. LeVay, *The Sexual Brain.*
16. D. H. Hamer, S. Hu, V. L. Magnuson, N. N. Hu and A. M. L. Pattatucci, 'A linkage between DNA markers . . .'.
17. Wilson, *Sociobiology*, p. 575.
18. See e.g. Mary Midgley, *The Ethical Primate.*
19. Furthermore, the objective sciences of biology and sociology omit a

crucial dimension of human existence, the subjective experience for each of us of our personal life histories. Despite recent attempts to redefine consciousness in terms of specific brain processes (e.g. by Francis Crick in *The Astonishing Hypothesis*), such subjectivity still lies outside the domain of the natural sciences at least, and may best be approached in novels and poetry. The unification of subjective and objective understandings of the nature and meaning of living processes – including human existence – if ever achievable, remains at best a distant goal.

20. H. G. Brunner, M. Nelen, X. O. Breakfield, H. H. Ropers and B. A. van Oost, 'Abnormal behavior . . .'.

21. Brunner, in Bock and Goode (eds), *Genetics of Criminal and Antisocial Behaviour*, pp. 164–7.

22. Olivier Cases *et al.*, 'Aggressive behavior . . .'.

23. Wade Roush, 'Conflict marks crime conference'.

24. Peter R. Breggin and Ginger Ross Breggin, 'A biomedical program for urban violence control . . .'; also evidence given by the Breggins to the National Institutes of Health panel on violence research.

25. Michael Rutter, Introduction to Bock and Goode (eds), *Genetics of Criminal and Antisocial Behaviour*, pp. 1–15.

26. See e.g. Leon Kamin, *The Science and Politics of IQ*.

27. Donald Mackenzie, *Statistics in Britain, 1865–1930*.

28. Richard J. Herrnstein and Charles Murray, *The Bell Curve*, p. 22.

29. L. S. Allen and R. A. Gorski, 'Sexual orientation and the size of the anterior commisure . . .'.

30. As reported by Chandler Burr, 'Homosexuality and biology'; and Richard C. Friedman and Jennifer Downey, 'Neurobiology and sexual orientation . . .'.

31. LeVay, 'A difference in hypothalamic structure . . .'.

32. Hamer *et al.*, 'A linkage between DNA markers . . .'.

33. Anne Fausto-Sterling, *Myths of Gender*.

34. David Fernbach, *The Spiral Path*.

35. V. H. Mark and F. R. Ervin, *Violence and the Brain*.

36. Quoted by Anne Moir and David Jessel in *A Mind to Crime*.

37. Peter D. Kramer, *Listening to Prozac*.

38. Steven Rose, *Molecules and Minds*.

39. Constance Holden, 'A cautionary genetic tale . . .'; Richard E. Chipkin, 'D2 receptor genes . . .'.

40. As quoted in the press release from the University of Wisconsin cited on page 275; see also Jerome Kagan, *The Nature of the Child*.

41. Robert Plomin, Michael J. Owen and Peter McGuffin, 'The genetic basis of complex human behaviors'.

42. Harriette C. Johnson, 'Violence and biology'; Stephen C. Maxon, 'Issues in the search for candidate genes in mice . . .'.
43. John C. Crabbe, John K. Belknap and Kari J. Buck, 'Genetic animal models of alcohol and drug abuse'.
44. Quoted in Moir and Jessel, *A Mind to Crime*.

11

Envoi: Making Biology Whole Again

The time has come for me to draw together the threads of argument which have been woven through the preceding ten chapters. My concern throughout has been to present a vision of living systems, living processes – indeed, life itself – from the perspective of modern biology. This vision is thoroughly materialist, recognizing the essentially historical nature of both our subject and our study of it, yet it offers an alternative to the current fashionable, deeply reductionist and deterministic accounts that dominate popular science writing, and indeed crowd the pages of some major scientific journals.

There is nothing greatly original here; I have not tried to – nor do I think it is necessary to – develop a new science of life, but rather to restate what to many practising biologists may seem the obvious, yet which has sadly been ignored or submerged by the tide of vulgar ultra-Darwinism and gung-ho biotechnological genetic sales talk that has threatened to engulf us. In doing so I have drawn not merely on present-day biochemical knowledge, but on two older traditions. One is still best described as dialectical, despite the almost irredeemable damage done to that term through the authoritarian aridities and monstrous social consequences of Soviet Marxism. The second is structuralist, though not applied with the same rejection of the historical perspective that characterizes the purist position adopted by some modern theorists. I have tried to acknowledge the personal and intellectual debts that I owe to both schools of thought in the Preface and in the notes to earlier chapters, but I remain only too aware that, as a biochemically trained neuroscientist, there are vast areas of biological knowledge with which I have only a modest acquaintance, notably the study of ecosystems and the intimate life histories of the 90 per cent plus of all species which get along quite well without brains or nervous systems. So much for disclaimers. Now to the argument, in the form of ten theses – biology's decalogue.

1. Our history shapes our knowledge

Our knowledge of the living world, like all other human knowledges, is always provisional, historically constrained. It is formed by the necessity both to interpret and to change the world. Confronted with the richly interacting complexity of the material world within which we are embedded, we abstract from it observations, processes, categories of objects (proteins, cells, organisms, species) to which we are inclined to grant the status of natural kinds. And we note the effects of changing the world by controlled intervention into these objects and processes – the art of experiment. These methods of acquiring knowledge are rule-bound. That is, we operate according to conventions about what constitutes an acceptable observation, experiment or interpretation which are profoundly shaped by the history of our subject, biology, the current social context, and our own ideological and intellectual preoccupations. Thus as scientists the representations we construct of the real, material world are required to conform to certain principles. Above all they must work, in that they must lead to consequences, whether experiments or technological artefacts, which do what is expected – predicted – of them. However, as I have argued, the fact that our experiments or technologies 'perform' in this sense does not in itself guarantee that they are based on true representations of the world.

In attempting to interpret and change the world, we often operate by analogy – by likening the process or object we are studying to another whose mechanism we understand more fully. Analogies, however, are hazardous tools. Often they are merely metaphor, and the likeness we imagine is poetic rather than exact. And often they are taken to imply homology – that is, that the process or object we are studying shares a common evolutionary descent with whatever it is we are analogizing it to. This is a powerful claim, and should not be made lightly.

2. *One world, many ways of knowing*

For any living phenomenon we observe and wish to interpret, there are many possible legitimate descriptions. In my fable of the five biologists and the jumping frog, there are within-level causal explanations, descriptions which locate the frog as part of a more complex ecosystem, and molecular, developmental and evolutionary accounts. These accounts cannot be collapsed into the 'one true' explanation in which the living phenomenon becomes 'nothing-but' a molecular assemblage, a genetic imperative, or whatever. It all depends on the purposes for which the explanation is required. To put it formally, we live in a material world which is an ontological unity, but which we approach with epistemological diversity. Biology, and the life processes it studies, will not conform to the proud manifesto of physics that the task of science is to reduce all accounts of the world to unitary theories of everything. Physics' claim will not work, and it is positively harmful to our understanding of living processes.

3. *Levels of organization*

Different scientific disciplines, from the social to the subatomic sciences, deal with different levels of organization of matter. The divisions between levels are, however, confused. In part they are ontological, and relate to scale and complexity, in which successive levels are nested one within another. Thus atoms are less complex than molecules, molecules than cells, cells than organisms, and organisms than populations and ecosystems. So at each level different organizing relations appear, and different types of description and explanation are required. Hence each level appears as a holon – integrating levels below it, but merely a subset of the levels above. In this sense, levels are fundamentally irreducible; ecology cannot be reduced to genetics, nor biochemistry to chemistry. However, to some extent – and this is where the confusion enters – the levels are epistemological, relating to different ways of knowing the world, each in turn the contingent product of its own discipline's history. The relationship between such epistemological levels (between biochemistry and physiology, say) is best described in the metaphor of translation. Thus the physiological language of contraction of the frog muscle can be translated into the biochemical language of the sliding filaments of actin and myosin.

Problems arise when one attempts to apply concepts and terms applicable at one level to phenomena on another level. Thus people may be gay or violent or schizophrenic or selfish, but brains or genes cannot be, in anything other than a metaphorical sense; equally, genes may replicate, but people cannot. But the power of metaphor is such that we always run the danger of confusing it with reality.

4. *It all depends . . .*

In living systems, causes are multiple and can be described at many different levels and in many different languages. Phenomena are always complex and richly interconnected. For example, the reasons why any individual contracts lung cancer or coronary heart disease will certainly relate to that person's unique genotype and developmental history, but also to such 'risk factors' as cigarette smoking, diet, work and living environment. What is required is to seek the *determining* cause – that is, the one that has the major effect on the system. For lung cancer, it is clearly cigarette smoking, and exploration of the molecular biology of the lungs, or of potentially 'predisposing' genes, becomes an arcane academic distraction, fostered in part by the tobacco lobby. By contrast, for Huntington's Disease the determining cause is clearly genetic and understanding the genetics and molecular biology may be the best strategy to alleviate or eliminate the disease. I have argued that, for such social concerns as urban violence, poverty and homelessness, to seek determining causes in genetics and biochemistry – as reductionist ideology attempts to do – is poor science and likely to lead to poor social prescriptions. Other conditions, such as the psychic anguish of schizophrenia or depression, remain contested zones, where crucial determinants may occur at several levels.

5. *Being and becoming*

Living organisms exist in four dimensions, the three of space and one of time, and cannot be 'read off' from the single dimension that constitutes the strand of DNA. Organisms are not empty phenotypes, related one-to-one to particular patterns of genes. Our lives form a developmental trajectory, or lifeline, stabilized by the operation of homeodynamic principles. This trajectory is not determined by our genes, nor partitioned into neatly dichotomous categories called nature and nurture. Rather, it is an autopoietic process, shaped by the interplay of specificity and plasticity. In so far as any aspect of life can be said to be 'in the genes', our genes provide the capacity for both specificity – a lifeline relatively impervious to developmental and environmental buffeting – and plasticity – the ability to respond appropriately to unpredictable environmental contingency, that is, to experience. This autopoietic interplay is in some senses captured by that old paradox of Xeno – the arrow shot at a target, which at any instant of time must be both somewhere and in transit to somewhere else. Reductionism ignores the paradox and freezes life at a moment of time. In attempting to capture its *being*, it loses its *becoming*, turning processes into reified objects. This is why reductionism always ends by impaling itself on a mythical dichotomy of materialist determinism and non-material free-will. Autopoiesis, self-construction, resolves these paradoxes.

6. *Stability through dynamics*

Organisms are open systems, far from thermodynamic equilibrium, in which continuity is provided by a constant flow of energy through them. Every molecule, every organelle, every cell, is in a constant state of flux, of formation, transformation and renewal. Dynamic stability of form persists, although every constituent of that form has been replaced. This stability, often maintained through oscillatory processes, depends on the capacity of complex interacting systems to self-organize, so as to maintain both short- and long-range order. Examples of such self-organization range from the self-assembly of proteins to form ribosomes or microtubules, and of lipids to form membranes, to the self-regulating metabolic web of enzymic interactions. In this view of living systems there are no master-molecules, no naked replicators controlling cellular events from within the screened-off tranquillity of the nuclear boardroom. Genes – lengths of DNA – are engaged in a continual metabolic interchange with other cellular components, a molecular democracy constrained by cellular organization, a cellular democracy constrained by the needs of the organism.

7. *Organism and environment interpenetrate*

Organisms are in constant interaction with their environment – put another way, organism and environment interpenetrate. That is, organisms actively select environments just as environments select organisms. Organisms move from unfavourable to favourable conditions; they absorb aspects of their environment – oxygen, food materials, metal ions – into themselves, and excrete waste – signal molecules or self-protective molecules. In doing so they constantly change their environments. The idea of a stable, unchanging environment, affected only by human and technological intervention, is a romantic fallacy. Like organisms, environments evolve and are homeodynamic rather than homeostatic.

8. *Structure constrains evolution*

Evolutionary change occurs as a result of the continued intersection of lifeline trajectories with changing environments. Such change occurs at many levels from the molecular to the species. The prime mechanism of this change, although not the only one, is natural selection, and it too operates at many levels, from the individual gene to the population. The replicative mechanisms provided by the cellular machinery enabling identical copies of DNA molecules to be synthesized of course mediate all of these selective mechanisms. There are, however, constraints on these selective processes. First, not all change is selectively adaptive: some may be contingent and essentially neutral in its effect. Second, because of the extent to which organisms select and modify environments, they are not simply the passive victims of selective processes, but play an active part in their own destiny. Third, evolution is not indefinitely flexible – not all that is possible is achievable. This is partly because living processes are in their essence only comprehensible in a historical context, and there are no such things in life as *de novo* engineering solutions to problems. The materials for evolutionary change are restricted to what is currently present. Opening certain evolutionary pathways closes others, and no evolutionary trajectory can move from a relatively high peak of fitness through a trough in pursuit of some distantly perceived higher peak. That is, selective processes cannot diminish an offspring's chances of life today in the hope that they will improve at some future time.

Furthermore, there are physical and chemical constraints on the structural possibilities available through evolution, from the rates of diffusion of dissolved gases, to the mechanical properties of the calcium phosphate of bones or the cellulose walls of plant cells. These limit cell size, body volumes, rates of movement, patterns of behaviour, and cannot be bypassed by any amount of genetic tinkering. Humans cannot be turned into angels by grafting onto us a genetic programme for wings, because no wing bone and muscle structure could achieve the lift to enable us to fly. (Instead, we possess, by courtesy of our evolutionary history, the cerebral, social and technical facilities to construct societies and machines enabling each and every one of us to fly, without the need for genetic change at all.) Whether there are deeper 'laws of form' than these, other than the patterns of self-organizing stability referred to above, remains unproven.

9. The past is the key to the present

It follows from this that organisms cannot predict the pattern of evolutionary change: they can only respond to present contingencies. And because all living organisms are simultaneously and continually responding to such contingencies, and in doing so changing the environment both for themselves and for others, evolutionary change can do nothing other than track a continually moving and inherently unpredictable target. The odds are always changing, at all levels from the molecular through the individual to the population and species. That is why evolution is indeed the 'law of higgledy-piggledy', and why nothing in biology makes sense except in the light of history.

10. Life constructs its own future

Thus for humans, as for all other living organisms, the future is radically unpredictable. This means that we have the ability to construct our own futures, albeit in circumstances not of our own choosing. And it is therefore our biology that makes us free.

Bibliography

Allen, Garland E. (1979) *Thomas Hunt Morgan: The Man and His Science*, Princeton University Press.

Allen, L. S. and Gorski, R. A. (1992) 'Sexual orientation and the size of the anterior commisure in the human brain', *Proceedings of the National Academy of Sciences of the USA*, 85, 7199–202.

Baker, Robin (1996) *Sperm Wars: Infidelity, Sexual Conflict and Other Bedroom Battles*, Fourth Estate.

Ball, Philip (1995) 'Spheres of influence', *New Scientist*, 2 Dec., pp. 42–4.

Barash, David (1981) *Sociobiology: The Whisperings Within*, Souvenir Press.

Behe, Michael (1996) *Darwin's Black Box*, Simon & Schuster.

Birke, Lynda (1994) *Feminism, Animals and Science: The Naming of the Shrew*, Open University Press.

Blakemore, C. B. and van Sluyters, R. C. (1974) 'Reversal of the physiological effects of monocular deprivation in kittens: Further evidence for a sensitive period', *Journal of Physiology*, 237, 195–216.

Bock, Gregory and Goode, Jamie (eds) (1996) *Genetics of Criminal and Antisocial Behaviour*, Wiley.

Bonner, John Tyler (1974) *On Development: The Biology of Form*, Harvard University Press.

Bonner, John Tyler (1988) *The Evolution of Complexity by Means of Natural Selection*, Princeton University Press.

Borges, Jorge Luis (1965) 'Funes the memorious', in *Fictions*, Calder, pp. 97–105.

Bouchard, Thomas (1997) 'Experience producing drive theory: Low genes drive experience and shape personality', *Acta Paediatrica*, Suppl. 422, 60–64. *Acta paediatrica*, in the press.

Bouchard, T. J., Segal, N. L. and Lykken, D. T. (1990) 'Genetic and environmental influences on special mental abilities in a sample of twins reared apart', *Acta genetica gemellologica*, 39, 193–206.

Bowler, Peter J. (1989) *The Mendelian Revolution*, Athlone Press.

Breggin, Peter R. and Breggin, Ginger Ross (1994) 'A biomedical program for urban violence control in the US; the dangers of psychiatric social control', Center for the Study of Psychiatry (mimeo).

Brunner, H. G., Nelen, M., Breakfield, X. O., Ropers, H. H. and van Oost, B.A. (1993) 'Abnormal behavior associated with a point mutation in the structural gene for monoamine oxidase A', *Science*, **262**, 578–80.

Budiansky, Stephen (1995) *Nature's Keepers*, Weidenfeld & Nicolson.

Buffon, Comte de (1774–9) *De la manière d'étudier et de traiter l'histoire naturelle*, Paris.

Bukharin, Nikolai *et al.* (1971) *Science at the Crossroads*, Cass [1st edn 1932].

Burkhardt, Richard W. (1977) *The Spirit of System: Lamarck and Evolutionary Biology*, Harvard University Press.

Burr, Chandler (1993) 'Homosexuality and biology', *Atlantic Monthly*, **271** (3), 47–65.

Byrne, R. W. (1994) 'The evolution of intelligence', in *Behaviour and Evolution*, ed. Peter J.B. Slater and Timothy R. Halliday, Cambridge University Press, pp. 223–65.

Cairns-Smith, A. Graham (1985) *Seven Clues to the Origin of Life: A Scientific Detective Story*, Cambridge University Press.

Cannon, Walter (1932) *The Wisdom of the Body*, Kegan Paul [reprinted 1947].

Cases, Olivier *et al.* (1995) 'Aggressive behavior and altered amounts of brain serotonin and norepinephrine in mice lacking MAOA', *Science*, **268**, 1763–8.

Chipkin, Richard E. (1994) 'D2 receptor genes – the cause or consequence of substance abuse?', *Trends in Neuroscience*, **17**, 50.

Cohen, David B. (1994) *Out of the Blue: Depression and Human Nature*, Simon & Schuster.

Concar, David (1995) 'Sex and the symmetrical body', *New Scientist*, 22 April, pp. 40–44.

Coon, Carleton S. (1963) *The Origin of Races*, Jonathan Cape.

Coveney, Peter and Highfield, Roger (1995) *Frontiers of Complexity: The Search for Order in a Chaotic World*, Faber & Faber.

Crabbe, John C., Belknap, John K. and Buck, Kari J. (1994) 'Genetic animal models of alcohol and drug abuse', *Science*, **264**, 1715–23.

Crick, Francis H. C. (1958) 'On protein synthesis', *Symposia of the Society for Experimental Biology*, **12**, 138–63.

Crick, Francis H. C. (1981) *Life Itself*, Simon & Schuster.

Crick, Francis H. C. (1989) 'Neural Edelmanism', *Trends in Neuroscience*, **12**, 240–42.

Crick, Francis (1994) *The Astonishing Hypothesis*, Simon & Schuster.

Dawkins, Richard (1976) *The Selfish Gene*, Oxford University Press.

Dawkins, Richard (1982) *The Extended Phenotype*, Freeman.

Dawkins, Richard (1986) *The Blind Watchmaker*, Longman.

Dawkins, Richard (1986) 'Sociobiology: The new storm in a teacup', in *Science and Beyond*, ed. Steven Rose and Lisa Appignanesi, Basil Blackwell, pp. 73–4.

Dawkins, Richard (1995) *River out of Eden*, Weidenfeld & Nicolson.

Delgado, José (1971) *Physical Control of the Mind: Towards a Psychocivilised Society*, Harper & Row.

Dennett, Daniel C. (1995) *Darwin's Dangerous Idea: Evolution and the Meanings of Life*, Allen Lane.

Desmond, Adrian and Moore, James (1991) *Darwin*, Michael Joseph.

Dickens, Charles (1969) *Hard Times*, Penguin, pp. 48, 126 [1st edn 1854].

Dobzhansky, Theodosius (1973) 'Nothing in biology makes sense except in the light of evolution', *American Biology Teacher*, 35, 125–9.

Drug and Chemical Evaluation Section (1995) 'Methylphenidate: A background paper', US Department of Justice Drug Enforcement Administration, Washington DC.

Dunbar, Robin (1995) *The Trouble with Science*, Faber & Faber.

Edelman, Gerald (1987) *Neural Darwinism: The Theory of Neuronal Group Selection*, Basic Books.

Edelman, Gerald (1988) *Topobiology: An Introduction to Molecular Embryology*, Basic Books.

Edelman, Gerald (1989) *The Remembered Present: A Biological Theory of Consciousness*, Basic Books.

Edelman, Gerald (1992) *Bright Air, Brilliant Fire: On the Matter of the Mind*, Basic Books.

Elena, Santiago F., Cooper, Vaughn S. and Lenski, Richard E. (1996) 'Punctuated evolution caused by selection of rare beneficial mutations', *Science*, **272**, 1802–4.

Fausto-Sterling, Anne (1992) *Myths of Gender: Biological Theories about Women and Men*, Basic Books.

Fernbach, David (1981) *The Spiral Path*, Gay Men's Press.

Feyerabend, Paul (1987) *Farewell to Reason*, Verso.

Feyerabend, Paul (1995) *Killing Time*, Chicago University Press.

Friedman, Richard C. and Downey, Jennifer (1993) 'Neurobiology and sexual orientation: Current relationships', *Journal of Neuropsychiatry and Clinical Neurosciences*, 5, 131–53.

Galanter, M. (ed.) (1993) *Recent Developments in Alcoholism*, Plenum.

Garcia-Bellido, Antonio (1994) 'How organisms are put together', *European Review*, 2, 15–21.

George, Wilma (1964) *Biologist Philosopher: A Study of the Life and Writings of Alfred Russel Wallace*, Abelard-Schuman.

Goldbeter, Albert (1996) *Biochemical Oscillations and Cellular Rhythms: The Molecular Bases of Periodic and Chaotic Behaviour*, Cambridge University Press.

Goodwin, Brian (1963) *Temporal Organization in Cells*, Academic Press.

Goodwin, Brian (1994) *How the Leopard Changed Its Spots*, Weidenfeld & Nicolson.

Goodwin, Brian and Dawkins, Richard (1995) 'What is an organism? A discussion', in *Behavioral Design*, Perspectives in Ethology, Vol. 11, ed. N.S. Thompson, Plenum, pp. 47–60.

Goodwin, Trevor W. (1987) *A History of the Biochemical Society 1911–1986*, The Biochemical Society (London).

Gould, Stephen Jay (1977) *Ontogeny and Phylogeny*, Belknap Press (Harvard).

Gould, Stephen Jay (1981) *The Mismeasure of Man*, Norton.

Gould, Stephen Jay (1989) *Wonderful Life: The Burgess Shale and the Nature of History*, Penguin.

Gould, Stephen J. and Eldredge, Niles (1977) 'Punctuated equilibria: The tempo and mode of evolution reconsidered', *Paleobiology*, 3, 110–27.

Gould, Stephen J. and Lewontin, Richard C. (1979) 'The spandrels of San Marco and the Panglossian paradigm: A critique of the adaptationist programme', *Proceedings of the Royal Society B*, **205**, 581–98.

Government White Paper (1993) *Realizing Our Potential: Strategy for Science, Engineering and Technology*, HMSO.

Hacking, Ian (1995) *Rewriting the Soul: Multiple Personality and the Science of Memory*, Princeton University Press.

Haken, Hermann, Karlqvist, Anders and Svedin, Uno (eds) (1993) *The Machine as Metaphor and Tool*, Springer-Verlag.

Haldane, J. B. S. (1946) 'The interaction of nature and nurture', *Annals of Eugenics*, **13**, 197–205.

Haldane, J. B. S. (1968) 'The origin of life', in *Science and Life*, Pemberton Publishing, pp. 1–11 [originally published in 1929].

Haldane, J. B. S. (1985) *On Being the Right Size and Other Essays*, Oxford University Press, p. 2.

Hales, Stephen (1961) *Vegetable Staticks*, Macdonald [1st edn 1727].

Hamer, Dean and Copeland, P. (1994) *The Science of Desire: The Search for the Gay Gene and the Biology of Behavior*, Simon & Schuster.

Hamer, Dean H., Hu, S., Magnuson, V. L., Hu, N. N. and Pattatucci, A. M. L. (1993) 'A linkage between DNA markers on the X chromosome and male sexual orientation', *Science*, **261**, 321–7.

Hamilton, William D. (1964) 'The genetical evolution of social behaviour, I and II', *Journal of Theoretical Biology*, 7, 1–32.

Haraway, Donna (1976) *Crystals, Fabrics and Fields*, Yale University Press.

Haraway, Donna (1989) *Primate Visions: Gender, Race and Nature in the World of Modern Science*, Routledge.

Hawking, Stephen and Penrose, Roger (1996) *The Nature of Space and Time*, Princeton University Press.

Herrnstein, Richard J. and Murray, Charles (1994) *The Bell Curve*, The Free Press.

Hess, Benno and Mikhailov, Alexander (1994) 'Self-organisation in living cells', *Science*, **264**, 223–4.

Hirsch, Jerry (1990) 'A nemesis for heritability estimation', *Behavioral and Brain Sciences*, **13**, 137–8.

Hirschleifer, J. (1977) 'Economics from a biological viewpoint', *Journal of Law and Economics*, **20**, 1–52.

Holden, Constance (1994) 'A cautionary genetic tale: The sobering story of D2', *Science*, **264**, 1696–7.

Hopkins, Frederick Gowland (1913) 'The dynamic side of biochemistry', *Nature*, **92**, 213–23 [the epigraph to Chapter 6 is from p. 220].

Hoyle, Fred and Wickramasinghe, Chandra (1978) *Lifecloud: The Origin of Life in the Universe*, Dent.

Hubbard, Ruth and Lewontin, Richard C. (1996) 'Pitfalls of genetic testing', *New England Journal of Medicine*, **334**, 1192–4.

Hubby, J. L. and Lewontin, Richard C. (1966) 'A molecular approach to the study of genic heterozygosity in natural populations. I The number of alleles at different loci in *Drosophila pseudoobscura*', *Genetics*, **45**, 97–104.

Jacob, François (1974) *The Logic of Living Systems*, Allen Lane.

Janov, Arthur (1990) *The New Primal Scream*, Sphere.

Johnson, Harriette C. (1996) 'Violence and biology: A review of the literature', *Families in Societies: The Journal of Contemporary Human Services*, 77, 3–17.

Jones, Steve (1996) *In the Blood: God, Genes and Destiny*, HarperCollins.

Kacser, Henry and Burns, J. A. (1979) 'Molecular democracy: Who shares the controls?', *Biochemical Society Transactions*, 7, 1149–60.

Kagan, Jerome (1994) *The Nature of the Child*, Basic Books.

Kamin, Leon (1974) *The Science and Politics of IQ*, Laurence Erlbaum.

Kauffman, Stuart (1995) *At Home in the Universe: The Search for Laws of Complexity*, Viking.

Kauffman, Stuart (1996) 'Even peptides do it', *Nature*, **382**, 496–7.

Kawai, M. (1965) 'Newly acquired precultural behaviour of the natural troop of Japanese monkeys on Koshima islet', *Primates*, **6**, 1–30.

Kay, Lily E. (1993) *The Molecular Vision of Life: Caltech, the Rockefeller Foundation and the Rise of the New Biology*, Oxford University Press.

Keller, Evelyn Fox (1983) *A Feeling for the Organism: The Life and Work of Barbara McClintock*, Freeman.

Kettlewell, H. B. D. (1955) 'Selection experiments on industrial melanism in the lepidoptera', *Heredity*, **9**, 323–42.

Khakina, L. N. (1992) *Concepts of Symbiogenesis: A Historical and Critical Study of the Research of Russian Botanists*, Yale University Press.

Kitcher, Philip (1985) *Vaulting Ambition*, MIT Press.

Koestler, Arthur (1964) *The Act of Creation*, Hutchinson.

Koestler, Arthur and Smythies, J. R. (1969) *Beyond Reductionism: New Perspectives in the Life Sciences*, Hutchinson.

Kohn, Marek (1995) *The Race Gallery*, Jonathan Cape.

Koshland, Daniel (1989) 'Editorial', *Science*, **246**, 189.

Kramer, Peter D. (1994) *Listening to Prozac,* Fourth Estate.

Kuhn, Thomas S. (1962) *The Structure of Scientific Revolutions*, Chicago University Press.

Kuper, Adam (1994) *The Chosen Primate: Human Nature and Cultural Diversity*, Harvard University Press.

Lakatos, Imre and Musgrave, Alan (1970) *Criticism and the Growth of Knowledge*, Cambridge University Press.

Latour, Bruno (1987) *Science in Action*, Open University Press.

Latour, Bruno (1993) *The Pasteurization of France*, Harvard University Press.

Latour, Bruno and Woolgar, Steve (1979) *Laboratory Life: The Social Construction of Scientific Facts*, Sage.

LeVay, Simon (1991) 'A difference in hypothalamic structure between heterosexual and homosexual men', *Science*, **253**, 1034–7.

LeVay, Simon (1993) *The Sexual Brain*, MIT Press.

Levi Montalcini, Rita (1988) *In Praise of Imperfection: My Life and Work*, Basic Books.

Lewontin, Richard C. (1974) *The Genetic Basis of Evolutionary Change*, Columbia University Press.

Lewontin, Richard C. (1991) *Biology as Ideology*, Harper.

Lewontin, Richard C. (1996) *Human Diversity*, Freeman.

Lewontin, Richard C. and Levins, Richard (1976) 'The problem of Lysenkoism', in *The Radicalisation of Science*, ed. Hilary Rose and Steven Rose, Macmillan, pp. 32–64.

Loeb, Jacques (1964) *The Mechanistic Conception of Life*, Belknap Press [1st edn 1912].

Lovelock, James E. (1979) *Gaia: A New Look at Life on Earth*, Oxford University Press.

Ludmerer, K. M. (1972) *Genetics and American Society*, Johns Hopkins University Press.

Løvtrup, Sven (1987) *Darwinism: The Refutation of a Myth*, Croom Helm.

MacIntyre, Ferren and Estep, Kenneth W. (1993) 'Sperm competition and the persistence of genes for male homosexuality', *Biosystems*, 31, 223–33.

McKay, David S. *et al.* (1996) 'Search for past life on Mars: Possible relic biogenic activity in Martian meteorite ALH84001', *Science*, 273, 924–30.

MacKenzie, Donald (1981) *Statistics in Britain, 1865–1930: The Social Construction of Scientific Knowledge*, Edinburgh University Press.

Mackintosh, Nicholas J. (ed.) (1995) *Cyril Burt: Fraud or Framed?* Oxford University Press.

Mainstone, Rowland (1975) *Developments in Structural Form*, MIT Press.

Margulis, Lynn and West, Oona (1993) *Gaia and the Colonisation of Mars*, mimeo for GSA Today.

Mark, V. H. and Ervin, F. R. (1970) *Violence and the Brain*, Harper & Row.

Maturana, H. R. and Vanela, F. J. (1980) *Autopoiesis and Cognition: the realisation of the living*, Boston Reidal.

Maturana, H. R. and Vanela, F. J. (1998) *The tree of knowledge: the biological roots of human understanding*, Bonton, Shambhala.

Maxon, Stephen C. (1996) 'Issues in the search for candidate genes in mice as potential animal models of human aggression', in *Genetics of Criminal and Antisocial Behaviour*, ed. Gregory Bock and Jamie Goode, Wiley, pp. 21–30.

Mayr, Ernst (1982) *The Growth of Biological Thought*, Belknap Press.

Mayr, Ernst (1988) *Towards a New Philosophy of Biology*, Belknap Press.

Mayr, Ernst (1991) *One Long Argument*, Allen Lane.

Medawar, Peter (1967) *The Art of the Soluble*, Methuen.

Medawar, Peter (1984) 'A view from the left', *Nature*, 310, 255–6.

Midgley, Mary (1994) *The Ethical Primate*, Routledge.

Miller, Stanley L. (1953) 'A production of amino-acids under possible primitive Earth conditions', *Science*, 117, 528–9.

Moir, Anne and Jessel, David (1995), *A Mind to Crime*, Michael Joseph.

Monod, Jacques (1971) *Chance and Necessity*, Collins.

Moore, James (1979) *The Post-Darwinian Controversies*, Cambridge University Press.

Moore, N. (1987) *The Bird of Time*, pp. 124–5, quoted in the Open University course Ecology, S 328, Book 1, Open University Press, 1994.

Morell, Virginia (1996) 'Genes *v.* teams: Weighing group tactics in evolution', *Science*, 273, 739–40.

Olby, Robert (1974) *The Path to the Double Helix*, Macmillan.

Oparin, Alexander (1938) *The Origin of Life on the Earth*, Macmillan.

Oyama, Susan (1986) *The Ontogeny of Information*, Cambridge University Press.

Pauling, Linus (1968) 'Reflections on the new biology', *UCLA Law Review*, 15, 267–72.

Perrett, D. L., May, K. A. and Yoshikawa, S. (1994) 'Facial shape and judgements of female attractiveness', *Nature*, **368**, 239–42.

Perutz, Max (1986) 'A new view of Darwinism', *New Scientist*, 2 Oct., pp. 36–8.

Perutz, Max (1988) 'Reply, from Perutz, on reductionism', *Trends in Biochemical Science*, 13, 206.

Piaget, Jean (1979) *Behaviour and Evolution*, Routledge & Kegan Paul.

Plomin, Robert, Owen, Michael J. and McGuffin, Peter (1994) 'The genetic basis of complex human behaviors', *Science*, **264**, 1733–7.

Pope, Alexander (1731–35) 'Epistle I (to Viscount Cobham)', in *Moral Essays*, London.

Popper, Karl (1959) *The Logic of Scientific Discovery*, Hutchinson.

Popper, Karl (1986) 1st Medawar Lecture, The Royal Society (unpublished).

Provine, William B. (1973) *The Origins of Theoretical Population Genetics*, University of Chicago Press.

Purves, Dale (1994) *Neural Activity and the Growth of the Brain*, Cambridge University Press.

Rauschecker, Josef P. and Marler, Peter (eds) (1989) *Imprinting and Cortical Plasticity*, Wiley.

Ravetz, Jerome R. (1971) *Scientific Knowledge and Its Social Problems*, Oxford University Press.

Reiss, A. and Roth, J. (1993) *Understanding and Preventing Violence*, National Academy Press.

Ridley, Matt (1993) *The Red Queen: Sex and the Evolution of Human Nature*, Penguin.

Rose, Hilary (1994) *Love, Power and Knowledge*, Polity.

Rose, Hilary and Rose, Steven (eds) (1976) *The Radicalisation of Science*, Macmillan.

Rose, Steven (1986) *Molecules and Minds*, Open University Press.

Rose, Steven (1988) 'Reflections on reductionism', *Trends in Biochemical Science*, 13, 160–62.

Rose, Steven (1988) 'Steven Rose replies', *Trends in Biochemical Science*, 13, 379–80.

Rose, Steven (1991) *The Chemistry of Life*, Penguin [1st edn 1966].

Rose, Steven (1992) *The Making of Memory*, Bantam.

Rose, Steven (1992) 'So-called "formative causation": A hypothesis disconfirmed', *Biology Forum*, 85, 445–53.

Rose, Steven, Lewontin, Richard C. and Kamin, Leon (1984) *Not in Our Genes*, Penguin.

Ross, Andrew (ed.) (1996) *Science Wars*, Duke University Press.

Roush, Wade (1995) 'Conflict marks crime conference', *Science*, **269**,1808–9.

Rushton, J. Philippe (1995) *Race, Evolution and Behavior: A Life History Perspective*, Transaction Publishers.

Russell, Claire and Russell, W. M. S. (1968) *Violence, Monkeys and Man*, Macmillan.

Rutter, Michael (1996) 'Introduction', in *Genetics of Criminal and Antisocial Behaviour*, ed. Gregory Bock and Jamie Goode, Wiley, pp. 1–15.

Sacks, Oliver (1985) *The Man who Mistook His Wife for a Hat*, Duckworth.

Sagan, Dorion and Margulis, Lynn (1993) 'God, Gaia and Biophilia', in *Biophilia*, ed. E. O. Wilson and S. Kellert, Island Press, pp. 345–64.

Sartre, Jean-Paul (1948) *Existentialism and Humanism*, Methuen, pp. 28, 34, 45, 54 [original French edn 1946].

Saunders, Peter T. and Ho, Mae-Wan (1976, 1981) 'On the increase in complexity in evolution, I and II', *Journal of Theoretical Biology* **63**, 375–84 and **80**, 515–30.

Shapiro, James A. (1995) 'Adaptive mutation: Who's really who in the garden?', *Science*, **268**, 373–4.

Sheldrake, Rupert (1981) *A New Science of Life*, Blond & Briggs.

Sheldrake, Rupert (1992) 'An experimental test of the hypothesis of formative causation', *Biology Forum*, **85**, 431–43.

Sheldrake, Rupert (1992), 'Rose refuted', *Biology Forum*, **85**, 455–60.

Smith, John Maynard (1993) *Did Darwin Get it Right?* Penguin.

Smith, R. L. (ed.) (1984) *Sperm Competition and the Evolution of Animal Mating Systems*, Academic Press.

Spanier, Bonnie B. (1995) *Im/partial Science: Gender Ideology in Molecular Biology*, Indiana University Press.

Stent, Gunther (1978) *The Paradoxes of Progress*, Freeman.

Stepan, Nancy (1982) *The Idea of Race in Science*, Macmillan.

Tanksley, S. D. (1993) 'Mapping polygenes', *Annual Review of Genetics*, **27**, 205–33.

Teilhard de Chardin, Pierre (1969) *The Phenomenon of Man*, Harper & Row [French 1st edn 1955].

Thompson, D'Arcy W. (1961) *On Growth and Form*, Cambridge University Press [1st edn 1917].

Thorpe, William H. (1978) *Purpose in a World of Chance*, Oxford University Press.

Trivers, Robert (1971) 'The evolution of reciprocal altruism', *Quarterly Review of Biology*, **4**, 35–57.

Tumlinson, James H., Lewis, W. Joe and Vet, Louise E. M. (1993) 'How parasitic wasps find their hosts', *Scientific American*, **266**, 100–106.

Visalberghi, Elisabetta and Fragaszy, Dorothy M. (1996) 'Pedagogy and imitation in monkeys: Yes, no or maybe?', in *The Handbook of Education and Human Development*, ed. D. R. Olson and N. Torrance, Blackwell Scientific, pp. 277–301.

Waddington, Conrad H. (ed.) (1968, 1969, 1970, 1972) *Towards a Theoretical Biology*, Vols 1–4, Edinburgh University Press.

Watson, James (1986) 'Biology: A necessarily limitless vista', in *Science and Beyond*, ed. Steven Rose and Lisa Appignanesi, Basil Blackwell, pp. 19–25.

Watson, James D. and Crick, Francis H. C. (1953) 'Genetical implications of the structure of deoxyribonucleic acid', *Nature*, 171, 964–7.

Webster, Charles (1975) *The Great Instauration*, Duckworth.

Webster, Gerry and Goodwin, Brian (1996) *Form and Transformation: Generative and Relational Principles in Biology*, Cambridge University Press.

Weinberg, Steven (1993) *Dreams of a Final Theory*, Hutchinson Radius.

Weiner, Jonathan (1994) *The Beak of the Finch*, Vintage.

West, Geoffrey B., Brown, James H. and Enquist, Brian J. (1997) 'A general model for the origin of allometric scaling laws in biology', *Science*, **276**, 122–6.

Whyte, Lancelot Law (1965) *Internal Factors in Evolution*, George Braziller.

Williams, R. J. P. and Frausto da Silva, J. J. R. (1996) *The Natural Selection of the Chemical Elements*, Oxford University Press.

Willis, Sarah (1992) 'The influence of psychotherapy and depression on platelet imipramine and paroxetine binding', Thesis, Open University.

Wilmut, I., Schnieke, A. E., McWhir, J., Kind, A. J. and Campbell, K. H. S. (1997) 'Viable offspring derived from fetal and adult mammalian cells', *Nature*, 385, 810–13.

Wilson, Edward O. (1975) *Sociobiology: The New Synthesis*, Harvard University Press.

Wilson, Edward O. (1978) *On Human Nature*, Harvard University Press.

Wilson, Edward O. and Kellert, S. (eds) (1993) *Biophilia*, Island Press.

Wolpert, Lewis (1991) *The Triumph of the Embryo*, Oxford University Press.

Wright, Robert (1995) '20th century blues', *Time*, 28 Aug., p. 35.

Wynne Edwards, V.C. (1962) *Animal Dispersion in Relation to Social Behaviour*, Oliver & Boyd.

Zeki, Semir (1994) *A Vision of the Brain*, Blackwell Scientific.

Zuckerman, Solly (1932) *The Social Life of Monkeys and Apes*, Kegan Paul.

Index

Note: page numbers in italics refer to illustrations